Calcium and Magnesium Metabolism in Early Life

Calcium and Magnesium Metabolism in Early Life

Edited by
Reginald C. Tsang, M.B.B.S.
Gamble Professor of Neonatology
University of Cincinnati Medical Center
and Executive Director
Perinatal Research Institute
Children's Hospital Medical Center
Cincinnati, Ohio

CRC Press
Boca Raton Ann Arbor London Tokyo

Library of Congress Cataloging-in-Publication Data

Calcium and magnesium metabolism in early life / edited by Reginald C. Tsang.
p. cm.
Includes bibliographical references and index.
ISBN 0-8493-4613-4
1. Infants--Metabolism. 2. Calcium--Metabolism. 3. Magnesium--Metabolism. 4. Pregnancy--Nutritional aspects. I. Tsang, Reginald C.
[DNLM: 1. Calcium--metabolism. 2. Magnesium--metabolism. 3. Metabolism--in infancy & childhood. 4. Pregnancy--metabolism. QU 130 C144 1995]
RJ128.C35 1995
612.3′924--dc20
DNLM/DLC
for Library of Congress 94-21058
CIP

Direct all inquiries to CRC Press, Inc., 2000 Corporate Blvd., N.W., Boca Raton, Florida 33431.

International Standard Book Number 0-8493-4613-4
Library of Congress Card Number 94-21058
Printed in the United States of America 2 3 4 5 6 7 8 9 0
Printed on acid-free paper

PREFACE

The understanding of calcium and magnesium metabolism in early life is undergoing rapid development, introducing exciting new concepts of the physiology and pathophysiology of mineral metabolism that have emerged recently. Fetal neonatal and infant growth challenge the mineral homeostatic system in a remarkable manner, and the complex interplay of genetics, hormonal responses, nutrition and external physical stresses allow an interesting and complex array of interactions. The hormonal interactions at a cellular and molecular level, in response to other internal and external influences, at a time when there is a rapid change in bone and mineral needs, provides special opportunities for understanding the complexity of the developing organism. The mother who is pregnant or lactating provides another view of mineral metabolism, viewed from the organism that has to "sacrifice" her own mineral metabolism to advance the mineral requirements of the fetus and infant, respectively, allowing for another layer of complexity, which includes the physiology of the placenta and mammary glands as active transport sites for minerals. In this book we have compiled six chapters that span pregnancy, placenta, the lactating woman, the neonate, the infant, and early childhood beyond infancy, bringing together leading experts in perinatal mineral metabolism to provide a comprehensive and yet concise overview of the field as it stands today.

Dr. Leslie Myatt describes the adaptive changes that occur during pregnancy and the developing fetus. Maternal physiologic changes are described in addition to the dietary mineral requirements during pregnancy. Fascinating recent data on the association of calcium intake and hypertension during pregnancy are described, in addition to calcium metabolism and intracellular calcium in hypertension. Clinical trials related to the effects of calcium intake on pregnancy are reported in detail.

Drs. Husain and Mughal describe fetal calcium metabolism, particularly concentrating on the placental transport of calcium, including the mechanisms and control of transport. Placental transport of magnesium is described also in terms of mechanisms and regulation. Particularly interesting recent information regarding parathyroid hormone related peptide and its apparent function in regard to placental transport is described.

Drs. Koo and Mimouni's chapter on normal calcium and magnesium metabolism in preterm infants emphasizes the changes that occur in serum concentrations of these minerals and the clinical disturbances of hyper- or hypocalcemia or hyper- or hypomagnesemia, with emphasis on prevention and therapy. Changes in metabolism that occur during the preterm period are discussed particularly in regard to Vitamin D metabolism. Skeletal demineralization is discussed in reference to rickets/osteopenia of prematurity and its prevention and therapy; this common problem of prematurity appears to be predominantly a mineral deficiency disorder.

Drs. Mimouni and Koo describe the physiology of calcium and mineral metabolism in the neonatal period, particularly emphasizing the physiological changes in calcium, magnesium and phosphate. Blood mineral changes in normal and abnormal circumstances are described and the pathophysiologic background, clinical course, and management are also delineated. Mineral excess and deficiency states are described particularly in relationship to hormonal imbalances.

Drs. Kalkwarf and Specker describe the changes in calcium during the lactation and weaning period. Specifically, they describe the increases in dietary intake and reduction in renal calcium excretion that appear to be the major factors maintaining overall calcium economy during lactation. During the weaning period, surprisingly, there is an increase in fractional calcium absorption which appears to compensate at least in part for the loss in bone density that occurs during lactation. The authors describe the weight of evidence that indicates that lactation per se does not increase (it may even reduce) a woman's risk of osteoporotic fracture later in life. The authors describe magnesium status during lactation, which includes increases predominantly in dietary intake of magnesium and reduction in renal excretion of magnesium to maintain overall economy of magnesium.

Drs. Allen and Greer describe calcium and magnesium metabolism beyond infancy, discussing how genetic, hormonal and environmental factors affect calcium and magnesium metabolism. Calcium availability and conservation are described from intestinal and renal regulatory aspects. Disorders of calcium homeostasis include conditions with mineral deficiency and excess, and the authors particularly describe the relationships of these conditions with parathyroid hormone and Vitamin D.

Reginald C. Tsang

THE EDITOR

Reginald Tsang, M.B.B.S., is the David G. and Priscilla R. Gamble Professor of Neonatology and Professor of Pediatrics, Obstetrics and Gynecology of the University of Cincinnati. He is the Executive Director of the Perinatal Research Institute and Vice-Chairman for Research and University Affairs of the Children's Hospital of Cincinnati. Dr. Tsang is the past president of the American College of Nutrition. He received the Nutrition Award of the American Academy of Pediatrics in 1989. He is a reviewer for numerous journals of various subjects including nutrition, pediatrics, metabolism, endocrinology and developmental biology. He has been invited to speak in 30 states and 23 foreign countries on topics of nutrition, diabetes, calcium and vitamin D metabolism. Dr. Tsang has authored or co-authored more than 300 scientific papers and book chapters.

CONTRIBUTORS

David B. Allen, M.D.
Pediatric Endocrinology
University of Wisconsin School of Medicine
Madison, Wisconsin

Frank R. Greer, M.D.
Wisconsin Perinatal Center
University of Wisconsin
Madison, Wisconsin

Shahid M. Husain, M.B., Ch.B., M.R.C.P.
Department of Child Health
University of Manchester
St. Mary's Hospital
Manchester, England

Heidi J. Kalkwarf, Ph.D.
Division of Neonatology
Children's Hospital Medical Center
Cincinnati, Ohio

Winston W.K. Koo, M.B.B.S., F.R.A.C.P.
Pediatrics, Obstetrics and Gynecology
University of Tennessee
Memphis, Tennessee

Francis Mimouni, M.D.
Department of Pediatrics
Maimonides Medical Center
Brooklyn, New York

M. Zulficar Mughal, M.B., Ch.B., M.R.C.P., D.C.H.
Department of Child Health
University of Manchester
St. Mary's Hospital
Manchester, England

Leslie Myatt, Ph.D.
Department of Obstetrics and Gynecology
Perinatal Research Institute
University of Cincinnati College of Medicine
Cincinnati, Ohio

Bonny L. Specker, Ph.D.
Department of Pediatrics
University of Cincinnati Medical Center
Cincinnati, Ohio

Reginald C. Tsang, M.B.B.S.
Perinatal Research Institute
Children's Hospital Medical Center
Cincinnati, Ohio

TABLE OF CONTENTS

Chapter 1

Calcium Metabolism in Pregnancy

Leslie Myatt

CONTENTS

I. INTRODUCTION

Pregnancy presents a major challenge to calcium homeostasis in the female. This challenge is increased by the calcium requirements of the developing fetus and by the continuing calcium requirements of the mother herself, together with the need to prepare for the subsequent calcium demands of lactation. Calcium homeostasis is further challenged by the dramatic cardiovascular adaptations that occur with pregnancy and these adaptions are ultimately regulated by calcium. By term, the fetus accumulates up to 350 mg of calcium per day, with the total fetal calcium content at term being approximately 30 grams. This, however is less than 3% of maternal skeletal content. This gain of calcium by the fetus from the mother must be balanced by increased intestinal absorption and/or diminished renal excretion of calcium by the mother while simultaneously maintaining maternal skeletal calcium content. The three hormones, 1,25 dihydroxyvitamin D_3, parathyroid hormone and

0-8493-4613-4/95/$0.00+$.50

calcitonin interact at the kidney, intestine and bone to influence calcium loss, acquisition and storage to regulate calcium homeostasis. The purpose of this chapter is to describe the mechanisms which maintain calcium homeostasis during pregnancy and to outline some of the changes which may occur in calcium homeostasis in pathologic pregnancy such as those associated with pregnancy-induced hypertension and preeclampsia.

II. HORMONAL REGULATION OF CALCIUM HOMEOSTASIS

Parathyroid hormone (PTH), calcitonin and 1,25 dihydroxyvitamin D_3 (1,25 $[OH]_2D_3$) regulate calcium homeostasis and calcium flux into polarized cells.[1] A decrease in ionized calcium concentration in blood perfusing the parotid gland results in PTH secretion to maintain serum calcium. PTH inhibits bone formation by osteoblasts and stimulates osteoclast bone resorption, resulting in increased calcium and phosphorous in the circulation. PTH also increases renal calcium retention and stimulates renal 1-α-hydroxylase activity to increase renal $1,25(OH)_2D_3$ synthesis which, in turn, stimulates intestinal calcium uptake and synergizes with PTH to activate bone osteoclasts, consequently increasing blood ionized calcium. Elevated blood ionized calcium will inhibit PTH secretion and increase calcitonin secretion from the thyroid C cells. Calcitonin inhibits osteoclast activity and blocks calcium mobilization from bone. Small fluctuations in serum ionized calcium concentrations result in the action of these hormones to maintain calcium homeostasis as follows: they lower serum ionized calcium which leads to increased PTH secretion resulting in decreased bone formation, increased bone resorption, increased renal reabsorption of calcium, increased renal excretion of phosphorus (to prevent hyperphosphatemia), and increased intestinal absorption of calcium and phosphorous as a result of increased renal $1,25(OH)_2D_3$ synthesis. The resulting hypercalcemia will be limited by increased calcitonin secretion, but decreased PTH secretion, in response to increasing serum ionized calcium concentrations, which result in decreased bone resorption, increased renal calcium excretion and decreased intestinal calcium absorption.

III. THE PHYSIOLOGIC CHALLENGE OF PREGNANCY

Pregnancy presents one of the largest physiologic challenges the body faces to homeostatic mechanisms. Dramatic changes occur, particularly cardiovascular, where increased plasma volume and cardiac output are present, but vascular resistance decreases to maintain low blood pressure. Most of the physiologic changes of pregnancy promptly revert at delivery. During pregnancy most calcium accretion is into the fetal skeleton which, at term, contains approximately 30 grams of calcium.[2] At term, the placenta also appears to be calcified to a variable extent and contains approximately 1 gram of calcium.[3] To accommodate the need for fetal growth and development while also maintaining her own system in a homeostatic state, the pregnant woman obviously undergoes a series of physiologic adjustments. A variety of these changes are involved with calcium metabolism. These changes begin soon after pregnancy begins, although the majority of fetal accretion of calcium (80%) occurs in the third trimester, the period when maternal demand is greatest. Fractional intestinal absorption of calcium increases from 27% to as high as 50%[4] to accommodate the increased demand, but begins earlier than the time of increased fetal demand. There is little data at present on maternal-skeletal calcium turnover rates during pregnancy, but maternal bone calcium content appears to be preserved during pregnancy and may be increased early in pregnancy prior to fetal demand.[4] In calcium tolerance tests, Gertner et al[5] found increased calciuric and calcemic responses in pregnant women when compared to nonpregnant women which led them to conclude that pregnancy can be characterized as a state of physiologic hypercalciuria. It appears, however, that the increased renal blood flow and glomerular filtration rates of pregnancy are responsible for the increase in calcium excretion[6] as there is a close correlation between calcium excretion and creatinine clearance. Despite this increased urinary excretion of calcium, it is generally thought that a positive calcium balance exists in pregnancy.

A. DIETARY CALCIUM IN PREGNANCY

During pregnancy the RDA for calcium is 1200 mg per day, irrespective of age, whereas in nonpregnant individuals, the RDA is 1200 mg per day for women aged 11–24 years and 800 mg per day for older women[7]. Pregnant individuals under the age of 25 years, who have a low dietary calcium intake are recommended for supplementation with 600 mg calcium per day,[7] and older pregnant individuals who do not ordinarily consume an adequate diet, are recommended for supplementation with 250 mg per day. Although the current RDA for calcium in pregnancy in the United States is 1200 mg per day, it is possible that this is less than the amount required, and in a large number of cases, pregnant individuals probably do not receive the required amount. In a recent British study Schofield et al[8] showed that mean calcium intakes of less than 1000 mg per day were recorded in pregnant women. Problems in maintaining calcium homeostasis in pregnancy may arise in the adolescent population where individuals in the 10 to 17 year age group need a retention of 400 mg calcium per day for their own bone mineralization. If the fetal demands of pregnancy are then placed upon this group of individuals, the positive balance of calcium from a 1 gram per day intake will be insufficient.[9] In these individuals, therefore, an intake of up to 2 grams per day has been suggested. However, despite these apparent deficits in calcium intake and balance in such groups of patients, normal pregnancies do occur; therefore, adaptations in calcium absorption or bone demineralization may be occurring to accommodate both the fetal demand and the high urinary output of calcium in pregnancy.

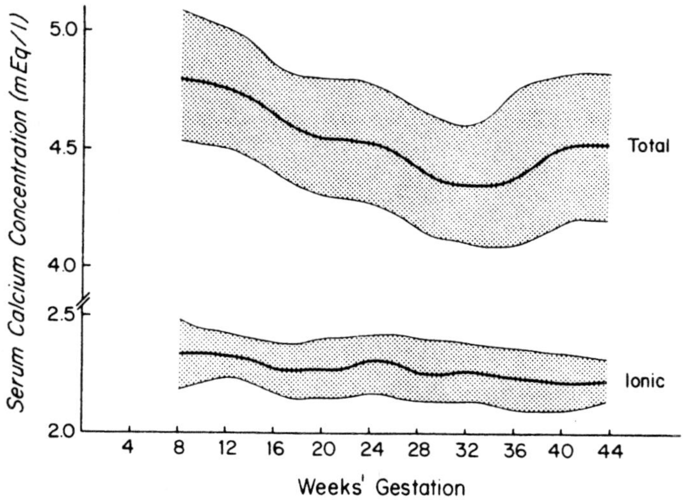

Figure 1 Maternal total and ionic calcium concentrations throughout gestation (mean ± 2.SD). (From Pitkin, R. M. and Gebhardt, M. P., *Am J Obstet Gynecol,* 127, 775, 1977. With permission).

B. SERUM CALCIUM IN PREGNANCY

Maternal serum total calcium concentration decreases progressively throughout pregnancy to a nadir in the middle of the third trimester and thereafter rises slightly until term[10] (Figure 1). The change in serum total calcium concentrations throughout pregnancy parallels the alterations in concentrations of serum albumin (Figure 2) which is the principal binding protein of calcium. In normal pregnancy, the concentration of serum ionized calcium remains constant throughout gestation[11] (Figure 1). These alterations in serum calcium throughout gestation appear to be independent of dietary intake in normal pregnancy; however, in situations where dietary intake of calcium is critically reduced, alterations in total and ionized calcium concentrations may occur (see below).

IV. CALCITROPIC HORMONES IN PREGNANCY

A. PARATHYROID HORMONE AND CALCITONIN

Due to the major alterations in maternal hormonal status which occur during pregnancy, the hormonal regulation of calcium metabolism in pregnancy has been extensively studied. Pregnancy has been traditionally viewed as a state of physiologic hyperparathyroidism[12] (Figure 3). Such an increase in parathyroid hormone secretion would be a logical maternal adjustment to expanding maternal extracellular fluid volume, increased urinary excretion and placental calcium transfer. Increased parathyroid hormone would stimulate bone resorption[1,13] and increase circulating calcium concentrations and intracellular calcium. As more specific assays have become available for parathyroid hormone, the view of pregnancy as a state of physiologic hyperparathyroidism has been

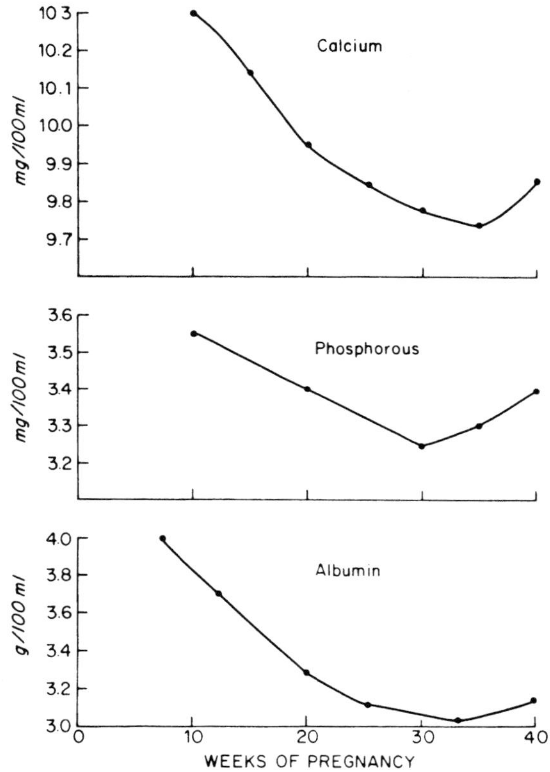

Figure 2 Maternal total calcium, phosphorous and albumin concentrations in pregnancy. (From Pitkin, R. M., *Am J Obstet Gynecol,* 121, 724, 1975. With permission).

challenged, with either no change or decreases in parathyroid hormone being reported.[14-18] There are equally conflicting reports for the influence of pregnancy on calcitonin secretion.[12,14,15] Calcitonin counteracts the hypercalcemic actions of PTH and 1,25$(OH)_2D_3$. Perhaps the calcitonin response is enhanced to protect the maternal skeleton during pregnancy as bone densitometric measurements have demonstrated no skeletal loss. Increased maternal calcitonin concentrations have been reported in the first[19] or second[20] trimesters of pregnancy and may be related to the calcium demand placed on the mother, although this has not been investigated.

B. VITAMIN D

The major circulating form of vitamin D, 25-hydroxy vitamin D(25-OHD_3), is generally unchanged by pregnancy. The concentrations reflect dietary intake and exposure to ultraviolet light and higher levels are found in summer and lower levels in winter. During pregnancy the biologically active form 1,25$(OH)_2D_3$ increases in concentration approximately twofold by term.[18,21-24]

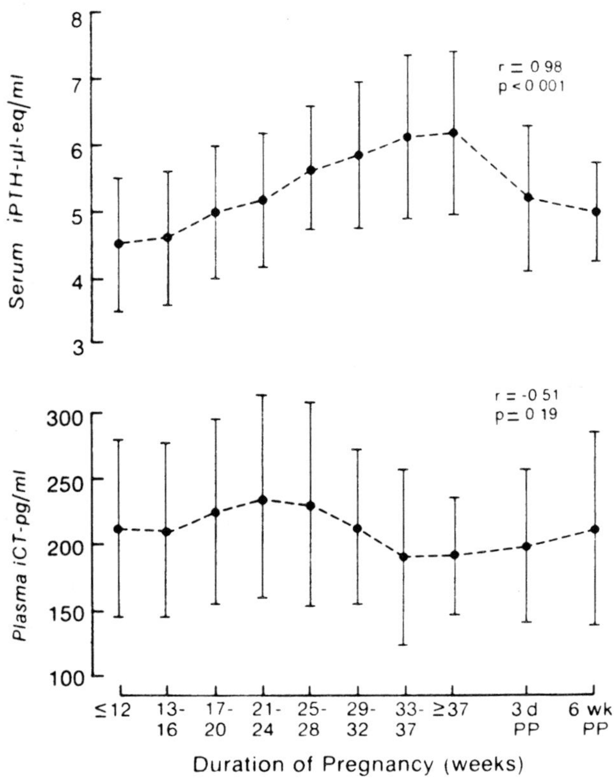

Figure 3 Maternal immunoreactive parathyroid hormone (iPTH) and calcitonin (iCT) concentrations in pregnancy and postpartum (PP) (Mean ± 2SD). (From Pitkin, R. M., Reynolds, W. A., Williams, G. A., Hargis, G. K., *Am J Obstet Gynecol,* 133, 781, 1979. With permission).

$1,25(OH)_2D_3$ can be synthesised in the placenta and/or decidua and in the fetal kidney. The source of increased maternal concentrations during pregnancy, however, is probably due to increased 1-α-hydroxylation of 25-OH vitamin D_3 in the maternal kidney.[25] The increase of serum $1,25(OH)_2D_3$ concentrations might be secondary to increases in serum PTH concentrations, PTH being a trophic regulator of renal 1-α-hydroxylase. The increased $1,25(OH)_2D_3$ is likely to be responsible for the increased intestinal absorption of calcium during pregnancy and probably helps maintains serum ionized calcium within a narrow range. Vitamin D is mainly synthesised in the skin, but also comes from dietary resources. The RDA for vitamin D is 10 μg (400 IU's) up to age 25 years and 5 μg (200 IU) afterwards.[26] During pregnancy the RDA is 10 μg (400 IU) with a 10 μg per day supplement being recommended for complete vegetarians and a 5 μg per day supplement for women with a low intake of vitamin D fortified milk, especially those in northern latitudes during winter.[7] Supplementation with vitamin D has been reported to have beneficial effects on birth weight, neonatal calcium homeostasis and maternal bone health in Indian populations in the United Kingdom where vitamin D deficiency is present.[27,28]

While the fetus accumulates 30 grams of calcium into its skeleton during pregnancy, this only represents less than 3 percent of the 1 kg of calcium stored and available in the adult female skeleton. Studies on maternal bone density during pregnancy have not found any significant effect on maternal bone content, particularly in long bones.[29-31] As yet, no definitive studies of maternal bone calcium turnover have been performed throughout gestation.

C. PLACENTAL TRANSFER OF CALCIUM AND CALCITROPIC HORMONES

The placenta moves calcium actively from mother to fetus against a concentration gradient so that by the end of gestation, the fetus is relatively hypercalcemic. This relative fetal hypercalcemia may be expected to result in decreased parathyroid secretion in the fetus and/or stimulation of fetal calcitonin secretion. However, parathyroid hormone concentrations in cord blood may not be altered[13], whereas fetal calcitonin secretion does appear to be increased. 25-hydroxy vitamin D_3 crosses the placenta from mother to fetus so that maternal and fetal levels correlate in a positive manner with fetal values being 20 to 30% below the maternal concentration. There is, however, no correlation of maternal and fetal concentrations of $1,25(OH)_2D_3$. Although it has been suggested[32] that the placenta can transfer $1,25(OH)_2D_3$, both placenta and the fetal kidney can produce it. As the fetal calcium is supplied and regulated by the placenta, the role of fetal $1,25(OH)_2D_3$ is unclear.

V. THE ASSOCIATION OF CALCIUM INTAKE WITH HYPERTENSION IN PREGNANCY

A. CALCIUM AND BLOOD PRESSURE

The intracellular calcium ion is a final effector in smooth muscle contractility. Therefore, mechanisms which alter intracellular calcium concentrations may alter vascular smooth muscle reactivity and uterine smooth muscle contractility. In nonpregnant subjects, epidemiologic studies have shown that restricted dietary calcium is associated with increased blood pressure.[33,34] Individuals with essential hypertension may have low serum ionized calcium concentrations.[35,36] Such findings have led to dietary calcium supplementation trials in normotensive and hypertensive populations. These studies have variable results. Calcium supplementation of up to 1300 mg of calcium per day for 12 weeks in normotensive males has been shown to modestly but significantly reduce mean arterial pressure.[37] Those patients with the greatest response to supplementation (greater or equal to 5 mmHg decrease in mean arterial blood pressure) were older and had higher mean arterial pressures, high serum PTH's and lower serum total calcium than did people who did not respond (less than

5 mmHg decrease in blood pressure). Other researchers[38,39] found that calcium supplementation had an antihypertensive effect in subgroups of hypertensive patients who had elevated PTH concentrations and low serum ionized calcium concentrations. Therefore, there does not appear to be a simplistic relationship between calcium supplementation and reduction in blood pressure, but there are other factors that may determine response to this treatment, particularly calcium regulating hormones.

B. CALCIUM AND HYPERTENSION IN PREGNANCY

During late pregnancy, episodic nocturnal cramping of the calf muscles is frequently experienced, thought to reflect transient hypocalcemia speculated to be related to excessive phosphate intake. However, treatment of leg cramps during pregnancy with calcium has proved disappointing.[40,41] There is, however, strong evidence for a link between calcium intake and the incidence of preeclampsia/eclampsia during pregnancy. Such an association has been suggested since the 1930's.[42] Hypertension during pregnancy may affect up to 10% of all pregnancies and this is the greatest cause of maternal morbidity and mortality. Hypertension which develops as a consequence of pregnancy and regresses postpartum is termed pregnancy-induced hypertension (PIH). This is defined as a systolic blood pressure of at least 140 mmHg, a diastolic of 90 mmHg, or an increase of systolic blood pressure of at least 30 mmHg or an increase in diastolic blood pressure of at least 15 mmHg over preexisting values and which occurs after the 20th week of gestation measured on two occasions at least 6 hours apart. The development of PIH with proteinuria, generalized edema or both, is termed preeclampsia, which is a well-defined clinical syndrome in pregnancy.

A high dietary calcium intake in pregnancy is associated with a relatively low incidence of eclampsia. The incidence of eclampsia is low in Ethiopia and rural Guatemala, two countries with low socioeconomic status and poor diets, where a high incidence of eclampsia might be expected. However, there is a high dietary intake of calcium (1075 and 1100 mg per day, respectively) in these two countries[43] due to the high calcium content of their stable diets of teff (a grain rich in calcium) and tortillas made with lime water. This is compared with dietary intakes of 240 to 350 mg calcium per day from countries such as Columbia, Thailand, Jamaica and India which have a high incidence of preeclampsia[43] or 673 mg per day, the intake reported in a low income U.S. population.[44] This epidemiologic evidence or linkage between reduced dietary calcium intake and hypertension in pregnancy is further supported by animal studies. In animal models of hypertension, dietary restriction of calcium is associated with hypertension in both nonpregnant[45,46] and pregnant[47] rats and sheep. Further, increase in calcium intake lowers blood pressure in such animals.[48,49] In twin pregnant sheep, the acute starvation or feeding of a low calcium diet (0.21% versus 0.71%) in the third trimester of pregnancy results in decreased serum calcium concentrations, increased systemic blood pressure, decreased uteroplacental blood flow and increased urine protein concentrations.[50,51] These results mimic the clinical presentation of preeclampsia in humans. Simple infusion of calcium gluconate was sufficient to return serum calcium concentrations to normal, reduce systemic blood pressure and increase uterine blood flow in these animals. This model for inducing hypertension only produces hypocalcemia in half of the sheep with twin pregnancies and is not effective in animals with singleton pregnancies. These findings point to the role that the fetus and its calcium requirements may play in the development of hypertension in pregnancy and also to the possibility that the differential response may represent differences in intestinal absorption or mobilisation from skeletal stores or overall retention of calcium. A further finding in these animals with reduced serum ionized calcium concentrations and increased blood pressure was that they also went into preterm labor and delivery, suggesting that both vascular and uterine muscle reactivity may have been altered.

VI. CALCIUM METABOLISM IN HYPERTENSIVE PREGNANCY

Overall, the data on mineral metabolism in hypertensive pregnancy are conflicting. This is probably a reflection of variable study designs and control groups. Significant decreases in serum immunoreactive PTH and ionized calcium and increased phosphorous concentrations were found in pregnant patients with essential hypertension compared to nonhypertensive pregnant controls.[52] Such differences may, however, have been due to the underlying hypertensive pregnancy. By comparison, Richards et al[53] reported no differences in serum ionized or total calcium concentrations between patients who had severe pregnancy-induced hypertension or were eclamptic and those who were normotensive, chronic hypertensives or had mild to moderate pregnancy-induced hypertensive. In a black South African population of severe pregnancy-induced hypertensive/eclamptic patients, total serum calcium concentrations were significantly lower than in normotensive controls, and the calculated ionized calcium concentration was significantly reduced in eclamptic patients compared to normotensive or severely hypertensive patients.[54] Waddell et al.[55] have found that at term, pregnant patients with pregnancy-induced hypertension had significantly lower serum total and ionized calcium concentrations than normotensive controls matched for maternal age and gestational age. The reductions were most marked in those patients with proteinuria. Cord blood calcium was also lower in the infants of the hypertensive patients, although not significantly lower than the controls. These findings are in agreement with reduced cord blood ionized calcium found in infants of essential hypertensive mothers compared with those of normotensive mothers.[52]

A. URINARY CALCIUM IN HYPERTENSIVE PREGNANCY

Although normal pregnancy is associated with increased urine calcium excretion, i.e., hypercalciuria,[6] pregnancy-induced hypertension or preeclampsia has been found to be associated with hypocalciuria. Reduced fractional urinary excretion of calcium was found in the third trimester of hypertensive patients compared to controls;[15] however, no differences in total serum calcium concentrations were found between the two groups of patients. The finding of hypocalciuria in preeclamptic patients or hypertensive patients with superimposed preeclampsia was suggested by Taufield et al[56] to be due to either reduced glomerular filtration rate or increased tubular reabsorption of calcium compared to the normotensive state. This hypocalciuria was not found in pregnant women with transient or chronic hypertension. However, it is not clear whether hypocalciuria is specific for preeclampsia or is a nonspecific manifestation of renal insufficiency associated with preeclampsia.[57] Screening for a low calcium/creatinine ratio in the late second to early third trimester of pregnancy was suggested as a test for predicting the development of preeclampsia.[58] However, Hutchesson et al.[59] in a prospective study involving 138 women were unable to demonstrate a reduction in urinary calcium excretion in preeclamptic women prior to the onset of hypertension and renal involvement.

VII. INTRACELLULAR CALCIUM IN HYPERTENSION

Although a linkage between reduced dietary intake of calcium and reduced serum calcium concentrations has been found, the mechanism that leads to increased vascular smooth muscle tone is still largely undefined. Hypertension appears to be associated with increased intracellular calcium concentrations which cause vascular contraction. Measurements of intracellular calcium in various blood cells have been made in hypertensive individuals. Although significantly increased platelet membrane-bound calcium concentrations[60] and platelet intracellular calcium concentrations were found in hypertensive subjects compared to normotensive controls and significantly correlated with systolic blood pressure,[61] lymphocyte intracellular calcium concentrations do not correlate with blood pressure.[62] Such differences in calcium handling may represent differences between cell types. In patients with pregnancy-induced hypertension, intracellular sodium, potassium and magnesium concentrations were not significantly different from gestational age-matched normotensive controls,[63] whereas erythrocyte intracellular calcium was significantly increased. In contrast, no differences in either basal or ADP-stimulated platelet intracellular calcium concentrations were found among nonpregnant patients, those in the third trimester who were normotensive or those who had pregnancy-induced hypertension or preeclampsia.[64] However, there was a tendency for reduction in response of platelet calcium concentration to 5-hydroxytryptamine stimulation which is most marked in preeclampsia. However, Kilby et al[65] reported significantly increased platelet intracellular calcium in proteinuric, hypertensive pregnant primigravida and a significant correlation between platelet intracellular calcium and systolic blood pressure. In a preliminary report of a longitudinal study throughout three trimesters of pregnancy in young black patients, Zemel et al[66] reported that although there was no difference in basal platelet intracellular calcium at any time, those patients who subsequently developed preeclampsia had an exaggerated increase in intracellular calcium in response to stimulation by arginine vasopressins. This response was maintained throughout the second and third trimesters of pregnancy and was a sensitive predictor of preeclampsia. As yet, there is still no agreement over intracellular calcium concentrations in hypertensive pregnancy.

VIII. MECHANISMS FOR REGULATING INTRACELLULAR CALCIUM HOMEOSTASIS

As previously stated, the increased intracellular calcium in hypertension is paradoxically accompanied by a decrease in extracellular calcium. The mechanisms responsible for this are as yet unclear (Figure 4). The action of PTH on cation transporters in the cell membrane and the activity of cation transporters themselves have received scrutiny of late. A reduction in the activity of Na^+/K^+ adenosine triphosphatase (ATPase) and a relative increase in the activity of Ca^{2+} ATPase have been reported in erythrocyte membranes of pregnancy-induced hypertensive women when compared to normotensive pregnant controls.[67] It is thought that the increase in Ca^{2+} ATPase may be a compensatory response by which the cell tries to extrude high concentrations of intracellular-free calcium which, in turn, may have arisen from increased sodium/calcium exchange by Na^+/Ca^{2+} ATPase in the presence of high intracellular sodium, itself due to a reduced Na^+/K^+ ATPase.[68] A direct increase in intracellular calcium may result from the action of PTH as a calcium ionophore in vascular smooth muscle.[69] PTH has also been shown to inhibit erythrocyte Na^+/K^+ ATPase which would indirectly increase intracellular calcium as described above.[63] Both serum PTH concentrations and concentrations of endoxin, a natriuretic substance that inhibits Na^+/K^+ ATPase, are higher in patients with pregnancy-induced hypertension when compared to controls, although the increase is only significant for endoxin.[63] However, in this group of patients, no differences were found in serum ionized calcium or magnesium concentrations. Increased PTH concentrations may result from decreased calcium intake and/or intestinal absorption or from increased urinary output of calcium. In calcium supplementation studies,[70,71] patients with the highest level of supplementation (2 grams of calcium per day) have the lowest serum PTH concentrations at term. Further support that the role of PTH may play in the control of blood pressure comes from the animal studies of Belizan et al[72] who found that rats fed a calcium-free diet developed increased blood pressure which could be prevented by parathyroidectomy. The parathyroidectomized calcium-free rats actually had lower blood pressures than control animals fed a normal calcium diet.

Although calcium is a final common effector for smooth muscle contractility and, hence, regulator of blood pressure, it may also have indirect effects by

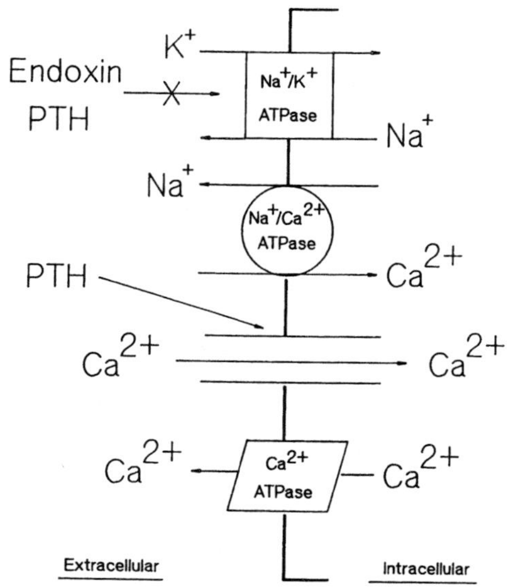

Figure 4 Mechanisms for maintenance of intracellular calcium homeostasis (From Myatt, L., in *Calcium Nutriture for Mothers and Children,* Eds. Tsang, R. C. and Mimouni, F., Raven Press, New York, 1992. With permission).

the regulation of production of other vasoactive agents. Extracellular calcium may stimulate prostacyclin synthesis from arachidonic acid by vascular tissue, the prostacyclin acting as a vasodilator.[73] It has also been suggested that extracellular calcium can regulate the synthesis of the novel endothelial cell-derived vasodilator, nitric oxide.[74] Nitric oxide synthesised from the amino acid L-arginine by the enzyme nitric oxide synthase in endothelium appears to have a major physiologic role in the control of vascular tone *in vivo*.[75]

IX. THE EFFECT OF DIETARY CALCIUM SUPPLEMENTATION IN PREGNANCY

Nutritional status throughout pregnancy has a major impact on the overall outcome of pregnancy including eclampsia and preterm labor.[76] Dietary supplementation studies have been performed during pregnancy with a series of nutrients including calcium. Although these studies led to reduction in the incidence of eclampsia, no clear knowledge of the particular supplement component responsible for this was found. When supplementing a lower-income pregnant population with a protein mineral supplement, Meritene, Osofsky et al[44] found a significantly lower increase in systolic and diastolic blood pressure in supplemented versus nonsupplemented patients. The supplemented group had a mean daily calcium intake of 1028 mg versus 673 mg in the nonsupplemented group. Our increased knowledge of the crucial role that calcium plays in regulation of vascular activity has led to dietary supplementation, although data from nonpregnant populations has yielded equivocal results. The greater challenges to calcium homeostasis during pregnancy and, hence, the necessity for greater intake during pregnancy perhaps suggest that the results from supplementation studies may be more clear-cut during pregnancy. Data from early studies have been highly encouraging and have now led to large placebo-controlled studies of calcium supplementation in pregnancy.

Belizan, Villar and co-workers[70,71] have studied the effect of calcium supplementation on blood pressure during pregnancy and on clinical outcome in two studies. In a study involving 36 urban Guatemalan women with single pregnancies, a calcium supplementation of 1 gram per day or 2 grams per day or a placebo was given from 15 weeks gestation until term.[70] Both calcium-supplemented groups had significantly lower diastolic blood pressures between 20 and 24 weeks gestation when compared with the placebo group. However, only patients receiving 2 grams of calcium per day had significantly lower blood pressures by the third trimester. The normal dietary intake of calcium in these patients was 700 mg per day, and no differences were observed in total blood calcium magnesium, phosphorous concentrations among the three study groups throughout the study. A further study by this group[71] involved 34 black patients from John Hopkins Hospital in Baltimore and 18 white patients from Rosario, Argentina, in a double-blind, randomized, placebo-controlled supplementation trial with 1.5 grams calcium per day after the 26th week of gestation. Subjects in the calcium-supplemented group had significantly lower systolic and diastolic blood pressures at term compared to the placebo group after adjustment for race and initial blood pressure and also a lower (4% versus 11.1%) incidence of pregnancy-induced hypertension, although this was not significant. A reduction in diastolic blood pressure (from 76.9 mmHg to 72.05 mmHg) with calcium supplementation found in this study would reduce the incidence of PIH from 10% to 4% assuming that blood pressure was normally distributed. The results of these two studies led the authors to believe that calcium supplementation had a dose-dependent effect on reducing blood pressure in pregnancy.

In another placebo-controlled study from India with supplementation of 375 mg per day of calcium and 1200 IU per day of vitamin D from 20 to 24 weeks gestation to term, a significant reduction in systolic and diastolic blood pressure by 36 weeks gestation was claimed, but there was only a slight significant reduction in the incidence of preeclampsia. Interestingly, by supplementing Japanese women with 156 mg calcium from mid-gestation to term, Kawasaki et al[77] found blunted vascular reactivity to infused angiotensin II without alteration in blood calcium measurements or blood pressure. Responses to infused angiotensin II are used as a screening test for development of pregnancy-induced hypertension in pregnant women[78] where exaggerated vascular activity to the infused angiotensin II is purportedly characteristic.

In a double-blind randomized placebo-controlled trial of 2 grams per day elemental calcium supplementation from 24 weeks gestation until term in a population of 106 native Indian women in Quito, Ecuador, where the usual dietary intake of calcium was 300 mg per day,[79] supplementation was found to be associated with a significantly decreased risk of pregnancy-induced hypertension (4.1%) when compared with a placebo group (27.9%).[80] Significant reductions in systolic and diastolic blood pressure were found over the duration of treatment, and a significantly greater increase in blood pressure throughout gestation was found in the placebo group. In the placebo group of patients, serum ionized calcium concentrations fell from 24 weeks gestation to 32 weeks gestation and remained low until delivery; the concentration increased again in the postpartum period. This progressive decline in serum ionized calcium throughout gestation in the placebo group with poor dietary calcium intake was presumably related to fetal demands. In the supplemented group, serum ionized calcium concentrations were significantly higher when compared to the placebo group and were increased by week 28 of gestation to concentrations seen in the nonpregnant postpartum period and remained at this level throughout gestation. Hence, calcium supplementation appears to be able to prevent the decline in serum ionized calcium seen in the placebo group and to maintain ionized calcium concentrations at the nonpregnant level. The duration of gestation was significantly increased in the calcium-supplemented group (39.3 weeks versus 38.7 weeks) which was probably partly responsible for the slightly, but not significantly higher, birth weight in the supplemented group (3097 grams versus 2832 grams).

The effect that normal dietary calcium intake may have on the outcome in such supplementation studies is illustrated by a similar trial in a predominantly black (94%) adolescent population in Baltimore.[81] In this study patients 17 years of age or less were randomized to either 2 grams per day elemental calcium supplementation or placebo in a double-blind randomised trial from 24 weeks gestation until term. The normal dietary intake of calcium in these patients was 1200 mg per day. There was a greater gestational age in the calcium group (39.2 weeks) when compared to the placebo group (37.9 weeks) which was probably responsible for the increase in birthweight of the calcium-supplemented group (3128 grams versus 2939 grams). A significantly lower incidence of preterm delivery (less than 37 weeks, 7.4% versus 21.1%), spontaneous preterm delivery (6.4% versus 17.9%) and low birthweight infants (9.6% versus 21.1%) was found in the calcium-supplemented group. There was, however, no significant reduction in the incidence of either pregnancy-induced hypertension or preeclampsia found in this study, although the rates were lower in the calcium-supplemented group. Such findings may reflect the fact that the basal calcium intake in this American adolescent pregnant group was higher than in the previously described Equadorian studies.

The diverse nature and small patient numbers involved in the clinical trials performed to date make it difficult to draw firm conclusions of the effect of dietary calcium supplementation in pregnancy. The different dietary, environmental, social and genetic factors found in each patient population make the data hard to interpret, but these factors also may be important determinants of response. However, supplementation of up to 2 grams elemental calcium per day may decrease reactivity of vascular smooth muscle and uterine smooth muscle needed to reduce blood pressure and uterine contractility. The effect of calcium supplementation may be profound in those patients with a deficiency of dietary calcium intake and correspondingly lower blood calcium concentrations.[80] The effect, however, may not be so clear-cut in those patients who have dietary calcium intakes in pregnancy which are close to the RDA.[7] In these latter patients, the endocrine regulation of calcium homeostasis may be more important than dietary calcium in regulation of vascular reactivity. However, particular patients who may be at theoretic higher risk for development of hypertension in pregnancy and for preterm labor include those with multiple gestation who have increased fetal demands for calcium and a greater incidence of preeclampsia and adolescent patients who have their own high ongoing demands for calcium in addition to those of pregnancy.

REFERENCES

1. Ross, R., Calcium regulating hormones, in *Neonatal and Fetal Medicine; Physiology and Pathology*, Polin, R.S. and Fox, W.W., Eds., S.B. Saunders, Philadelphia, 1991, 173.
2. Forbes, G. B., Calcium accumulation by the human fetus, *Pediatrics*, 57, 976, 1976.
3. Pitkin, R. M., Calcium metabolism in the pregnant and lactating female, in *Calcium Nutrition for Mothers and Children*, Tsang, R. C. and Mimouni, F., Eds., Carnation Nutriture Education Series *3* Carnation Co., Glendale 1, Raven Press, New York, 1992, 27.
4. Heaney, R. P., Skillman, T. G., Calcium metabolism in normal human pregnancy, *J Clin Endocrinol Metab*, 36, 661, 1971.
5. Gertner, J. M., Coustan, D. R., Kliger, A. S., Mallette, L. E., Ravin, N., Broadus, A. E., Pregnancy as state of physiologic absorptive hypercalciuria, *Am J Med*, 81, 451, 1986.
6. Howarth, A. T., Morgan, D. B., Payne, R. B., Urinary excretion of calcium in late pregnancy and its relation to creatinine clearance, *Am J Obstet Gynecol*, 129, 499, 1977.
7. Institute of Medicine. Committee on Nutritional Status during Pregnancy and Lactation. *Nutrition during pregnancy: I. Weight gain. II. Nutrient supplements*, Washington, DC, National Academy Press, 1990.
8. Schofield, C., Stewart, J., Wheeler, E., The diets of pregnant and post-pregnant women in different social groups in London and Edinburgh: calcium, iron retinol, ascorbic acid and folic acid, *Br J Nutr*, 62, 363, 1989.
9. Duggin, G. G., Lyneham, R. C., Dale, N. E., Evans, R. A., Tiller, D. J., Calcium balance in pregnancy, *Lancet*, 11, 926, 1974.
10. Pitkin, R. M., Calcium metabolism in pregnancy: a review, *Am J Obstet Gynecol*, 121, 724, 1975.
11. Pitkin, R. M., Gebhardt, M. P., Serum calcium concentrations in human pregnancy, *Am J Obstet Gynecol*, 127, 775, 1977.

12. Pitkin, R. M., Reynolds, W. A., Williams, G. A., Hargis, G. K., Calcium metabolism in pregnancy: a longitudinal study, *Am J Obstet Gynecol*, 133, 781, 1979.
13. Pitkin, R. M., Calcium metabolism in pregnancy and the perinatal period: a review, *Am J Obstet Gynecol*, 151, 99, 1985.
14. Whitehead, M., Lane, G., Young, O., Campbell, S., Abeyasekera, G., Hillyard, C. J., MacIntyre, I., Phang, K. G., Stevenson, J. C., Interrelationships of calcium-regulating hormones during normal pregnancy, *Br Med J*, 383, 10, 1981.
15. Pedersen, E. B., Johannesen, P., Kristensen, S., Rasmuesen A. B., Emmertsen, K., Moller J., Lauritsen, J. G., Wohlert, M., Calcium, parathyroid hormone and calcitonin in normal pregnancy and preeclampsia, *Gynecol Obstet Invest*, 18, 156, 1984.
16. Davis, O. K., Hawkins, D. S., Rubin L. P., Posillico, J. T., Brown, E. M., Schiff, I., Serum parathyroid hormone (PTH) in pregnant women determined by an immunoradiometric assay for intact PTH, *J Clin Endocrinol Metab*, 67, 850, 1988.
17. Salinas, M., Martinez, M. E., Catalan, P., Ordas, J., Navarro, P., Biochemical parameters of calcium metabolism in pregnant women: influence of milk supplementation, *Nutrition*, 3, 407, 1987.
18. Mimouni, F., Tsang, R. C., Hertzberg, V. S., Neumann, V., Ellis, K., Parathyroid hormone and calcitriol changes in normal and insulin-dependent diabetic pregnancies, *Obstet Gynecol*, 74, 49, 1989.
19. Stevenson, J. C., Hillyard, C., MacIntyre, I, Cooper, H., Whitehead, M.I., A physiological role for calcitonin: Protection of the maternal skeleton, *Lancet*, 2, 769, 1979.
20. Woluszczuk, W., Kovarik, J., Pavelka, P., Calcitonin in pregnancy women and in cord blood, *Gynecol Obstet Invest*, 12, 272, 1981.
21. Kumar, R., Cohen, W. R., Silva, P., Epstein, F. H., Elevated 1,25-dihydroxyvitamin D plasma levels in normal human pregnancy and lactation, *J Clin Invest* 4, 63, 342, 1979.
22. Bouillon, R., Van Assche, F. A., Van Baelen, H., Heyns, W., De Moor, P., Influence of the vitamin D-binding protein on the serum concentration of 1,25-dihydroxyvitamin D_3, *J Clin Invest*, 67, 589, 1981.
23. Reddy, G. S., Norman, A. W., Willis, D. M., Goltzman, D., Guyda H., Soloman, S., Phillips, D. R., Bishop, J. E., Mayer, E., Regulation of vitamin D metabolism in normal human pregnancy, *J Clin Endocrinol Metab*, 56, 363, 1979.
24. Markestad, T., Ulstein, M., Aksnes, L., Aarskog, D., Serum concentrations of vitamin D metabolites in vitamin D supplemented pregnant women. A longitudinal study, *Acta Obstet Gynecol Scand*, 65, 63, 1986.
25. Turner, M., Barre, P. E., Benjamin A., Goltzman, D., Gascon-Barre M., Does the maternal kidney contribute to the increased circulating 1,25-dihydroxyvitamin D concentrations during pregnancy? *Mineral Electrolyte Metab*, 14, 246, 1988.
26. National Research Council. Recommended Dietary Allowance, (10th ed.), Washington, DC, *National Academy Press*, 1989.
27. Brooke, O. G., Brown, I. R. F., Bone, C. D. M., Carter, N. D., Cleeve, H. J. W., Maxwell, J. D., Robinson, V. P., Winder, S. M., Vitamin D supplements in pregnant Asian women: effects on calcium status and fetal growth, *Br Med J*, 280, 751, 1980.

28. Marya, R. K., Rathee, S., Dua, V., Sangwan, K., Effect of vitamin D supplementation during pregnancy on foetal growth, *Indian J Med Res*, 88, 488, 1988.
29. Lamke, B., Brundin, J., Moberg, P., Changes of bone mineral content during pregnancy and lactation, *Acta Obstet Gynecol Scand*, 56, 217, 1977.
30. Christianson, C., Rodbro, P., Heinild, B., Unchanged total body calcium in normal human pregnancy, *Acta Obstet Gynecol Scand*, 55, 141, 1976.
31. Sowers, M. F., Crutchfield M., Jannausch, M., Updike S., Corton, G., A prospective evaluation of bone mineral change in pregnancy, *Obstet Gynecol*, 77, 841, 1991.
32. Devaskar, U. P., Ho, M., Devaskar, S. U., Tsang, R. C., 25-hydroxy- and Iα-dihydroxyvitamin D. Maternal-fetal relationship and the transfer of 1,25-dyhydroxyvitamin D_3 across the placenta in an ovine model, *Dev Pharmacol Ther*, 7, 213, 1984.
33. McCarron, D. A. Morris, C. D., Cole, C., Dietary calcium in human hypertension, *Science*, 217, 267, 1982.
34. Kesteloot, H., Joossens, J. V., Relationship of dietary sodium, potassium, calcium and magnesium with blood pressure, Belgian Interuniversity Research on Nutrition and Health, *Hypertension*, 12, 594, 1982.
35. McCarron, D. A., Low serum concentrations of ionized calcium in patients with hypertension., *N Engl J Med*, 307, 226, 1982.
36. Resnick, L. M., Uniformity and diversity of calcium metabolism in hypertension. A conceptual framework, *Am J Med*, 82, 16, 1987.
37. Lyle, R. M., Melby, C. L., Hyner, G. C., Metabolic differences between subjects whose blood pressure did or did not respond to oral calcium supplementation, *Am J Clin Nutr*, 47, 1030, 1988.
38. Grobbee, D. E., Hofman, A., Effect of calcium supplementation on diastolic blood pressure in young people with mild hypertension., *Lancet*, 2, 703, 1986.
39. Resnick, L. M., Nicholson, J. P., Laragh, J. H., Calcium metabolism in essential hypertension: relationship to altered renin system activity., *Fed Proc*, 45, 2739, 1986.
40. Hammar, M., Larsson, L., Tegler, L., Calcium treatment of leg cramps in pregnancy, Acta *Obstet Gynecol Scand*, 60, 345, 1981.
41. Hammar, M. G., Berg, G., Solheim, F., Larsson, L., Calcium and magnesium status in pregnant women. A comparison between treatment with calcium and vitamin C in pregnant women with leg cramps, *Int J Vitam Nutr Res*, 57, 179, 1987.
42. Theobald, G. W., Calcium and vitamins A and D on incidence of pregnancy-induced toxaemia, *Lancet*, 1, 1397, 1937.
43. Belizan, J. M., Villar, J., The relationship between calcium intake and edema proteinuria and hypertension gestosis: an hypothesis, *Am J Clin Nutr*, 33, 2202, 1980.
44. Osofsky, H. J., Relationship between prenatal medical and nutritional measures, pregnancy outcome and early infant development in an urban poverty setting, *Am J Obstet Gynecol*, 123, 682, 1975.
45. Furspan, P., Rinaldi, G. J., Hoffman, K., Bohr, D. F., Dietary calcium and cell membrane abnormality in genetic hypertension., *Hypertension*, 13, 727, 1989.
46. Hatton, D. C., Scrogin, K. E., Metz, J. A., McCarron, D. A., Dietary calcium alters blood pressure reactivity in spontaneously hypertensive rats, *Hypertension*, 13, 622, 1989.

47. Belizan, J. M., Pineda, O., Sainz, E., Meneridez, L. A., Villar, J., Rise of blood pressure in calcium-deprived pregnant rats, *Am J Obstet Gynecol*, 141, 163, 1981.
48. Ayachi, S., Increased dietary calcium lowers blood pressure in the spontaneously hypertensive rat, *Metabolism*, 12, 1234, 1979.
49. Peuler, J. D., Morgan, D. A., Mark, A. L., High calcium diet reduces blood pressure in Dahl-salt sensitive rats by neural mechanisms, *Hypertension*, 9, 159, 1987.
50. Prada, J., Ross, R., Clark, K., Hypertensive effects of starvation: role of Ca^{2+} in the pregnant sheep. Presented at the 34th Meeting of the Society for Gynecologic Investigation, 1987, abstract 62P.
51. Prada, J., Brockman, D., Moore, K, Clark, K., Hypocalcemia leads to pregnancy-induced hypertension in sheep. Presented at the 35th Meeting of the Society for Gynecologic Investigation, 1988, abstract 7.
52. Varner, M. W., Cruikshank, D. P., Pitkin, R. M., Calcium metabolism in the hypertensive mother, fetus and newborn infant, *Am J Obstet Gynecol*, 147, 762, 1983.
53. Richards, S. R., Nelson, D. M., Zuspan, F. P., Calcium levels in normal and hypertensive pregnant patients, *Am J Obstet Gynecol*, 149, 168, 1984.
54. Moodley, J., Rampersadh, S., Becker, P., Norman, R. J., O'Dondl, D., Serum calcium ion concentrations in eclampsia, *S Afr Med J*, 72, 382, 1987.
55. Waddell, B. J., Ross, R., Myatt, L., Disturbed fetal and maternal calcium and magnesium homeostasis in hypertensive pregnancy. Presented at VII World Congress of Hypertension in Pregnancy, 1990, abstract, page 269.
56. Taufield, P. A., Ales, K. L., Resnick, L. M., Druzin, M. L., Gertner, J. M., Laragh, J. H., Hypocalciuria in preeclampsia, *N Engl J Med*, 316, 715, 1987.
57. Graber, M. L., Moore, B., Hypocalciuria in preeclampsia, *N Engl J Med*, 317, 897, 1987.
58. Rodriguez, M. H., Masaki, D. I, Mestman, J., Kumar, D., Rude, R., Calcium/creatinine ratio and microalbuminuria in the prediction of preeclampsia, *Am J Obstet Gynecol*, 159, 1452, 1988.
59. Hutchesson, A. C. J., Macintosh, M. C., Duncan S. L. B., Forrest, A. R. W, Hypocalciuria and hypertension in pregnancy: a prospective study, *Clin and Exper Hyper in Pregnancy*, B9, 115, 1990.
60. Cooper, R., Lipowski, J., Ford, E., Shamsi, N., Feinberg, H., LeBreton, G., Increased membrane-bound calcium in platelets of hypertensive patients, *Hypertension*, 13, 139, 1989.
61. Pritchard K, Raine, A. G., Ashley, C. C., Castell, L. M., Somers, V., Osborn, C., Ledingham, J. G. G., Conway, J, Correlation of blood pressure in normotensive and hypertensive individuals with platelets but not lymphocyte intracellular free calcium concentrations, *Clin Sci*, 76, 631, 1989.
62. Lew, P. D., Favre, L., Waldovogel F. A., Vallotton, M. B., Cytosolic free calcium and intracellular calcium stores in neutrophils from hypertensive subjects, *Clin Sci*, 69, 227, 1985.
63. Sowers, J. R., Zemel, M. B., Bronsteen, R. A., Zemel, P. C., Walsh, M. F., Standley, P. R., Sokol, R. J., Erythrocyte cation metabolism in preeclampsia, *Am J Obstet Gynecol*, 161, 441, 1989.

64. Barr, S. M., Lees, K. R, Butters, L., O'Donnell, A., Rubin, P. C., Platelet intracellular free calcium concentration in normotensive and hypertensive pregnancies in the human, *Clin Sci*, 76, 67, 1989.
65. Kilby, M. D., Broughton-Pipkin F., Heptinstall, S., Symonds, E. M., A comparison of platelet cytosolic free calcium concentrations in both normotensive and hypertensive primigravidae. Presented at VII World Congress of Hypertension in Pregnancy, 1990, abstract, page 72.
66. Zemel, M. B., Zemel, P. C., Berry S., Norman, G., Kowakzyk, C., Sokol, R. J., Standley, P. R., Walsh, M. F., Sowers, J. R., Altered platelet calcium metabolism as an early predictor of increased peripheral vascular resistance and preeclampsia in urban black women, *N Engl J Med*, 323, 434, 1990.
67. Tranquili, A. L. Rabini, R. A., Cugini, A. M., Cester, N., Valensise H., Romanini, C., Bertoli, E., Mazzanti, L., Calcium and sodium transport in gestational hypertension, *Int J Cardiol*, 25, S53, 1989.
68. Friedman, S. M., Cellular ionic perturbations in hypertension, *J Hypertens*, 1, 190, 1983.
69. Schleiffer, R., Berthelot, A., Gairard, A., Effects of parathyroid extract on blood pressure and arterial contraction and on Ca exchange in isolated aorta in the rat, *Blood Vessels*, 16, 220, 1979.
70. Belizan, J. M., Villar, J., Zalazar, A., Rojas, J., Chan, D., Bryce, G. F., Preliminary evidence of the effect of calcium supplementation on blood pressure in normal pregnant women, *Am J Obstet Gynecol*, 146, 175, 1983.
71. Villar, J., Repke, J., Belizan, J. M., Pareja, G., Calcium supplementation reduces blood pressure during pregnancy: results of a randomized controlled clinical trial, *Obstet Gynecol*, 70, 317, 1987.
72. Belizan, J. M., Villar, J., Self, S., Pineda, O., Gonzalez, I., Sainz, E., The mediating role of the parathyroid gland in the effect of low calcium intake on blood pressure in the rat, *Arch Latinoam Nutr*, 34, 666, 1984.
73. Lopez-Jaramillo, P., Guarner, F., Moncada, S., Effects of calcium and parathyroid hormone on prostacyclin synthesis by vascular tissue, *Life Sci*, 40, 983, 1987.
74. Lopez-Jaramillo, P., Gonzalez, M. C., Palmer, R. M. J., Moncada, S., The crucial role of physiological Ca^{2+} concentrations in the production of endothelial nitric oxide and the control of vascular tone, *Br J Pharmacol*, 101, 489, 1990.
75. Rees, D. D., Palmer, R. M. J., Moncada, S., Role of endothelium-derived nitric oxide in the regulation of blood pressure, *Proc Natl Acad Sci USA*, 86, 3375, 1989.
76. Marya, R. K., Rathee, S., Manrow, M., Effect of calcium and vitamin D supplementation on toxaemia of pregnancy, *Gynecol Obstet Invest*, 24, 38, 1987.
77. Kawasaki, N., Matsui, K., Ito, M., Nakamura, T., Yoshimura, T., Ushijima, H., Maeyama, M., Effect of calcium supplementation on the vascular sensitivity to angiotensin II in pregnant women, *Am J Obstet Gynecol*, 153, 576, 1985.
78. Gant, N. F., Daley, G. L., Chand, S., Whalley, P. J., MacDonald P. C., A study of angiotensin II pressor response through primigravid pregnancy, *J Clin Invest*, 52, 2682, 1973.

79. Lopez-Jaramillo, P., Narvaez, M., Yepez, R., Effect of calcium supplementation on the vascular sensitivity to angiotensin II in pregnant women, *Am J Obstet Gynecol*, 156, 261, 1987.
80. Lopez-Jaramillo, P., Narvaez, M., Weigel, R. M., Yepez, R., Calcium supplementation reduced the risk of pregnancy-induced hypertension in an Andes population, *Br J Obstet Gynaecol*, 96, 648, 1989.
81. Villar, J., Repke, J., Calcium supplementation during pregnancy may reduce preterm delivery in high-risk populations, *Am J Obstet Gynecol*, 163, 1124, 1990.

Chapter 2

Fetal Calcium and Magnesium Metabolism

Shahid M. Husain and M. Zulficar Mughal

CONTENTS

0-8493-4613-4/95/$0.00+$.50

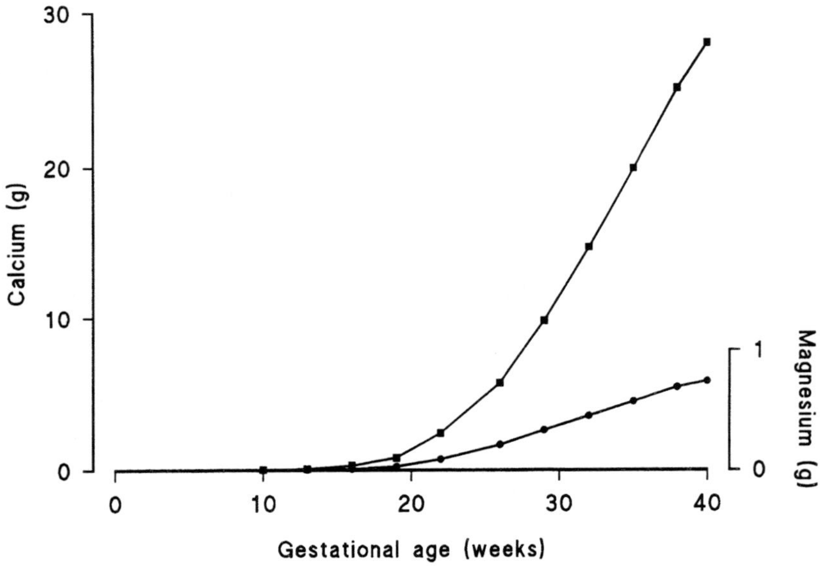

Figure 1 Total content of calcium (■) and magnesium (●) in the human fetus with increasing gestational age. (Adapted from Reference 1.)

I. MATERNOFETAL CALCIUM AND MAGNESIUM RELATIONSHIPS

During the course of normal human pregnancy the fetus accumulates approximately 28 g of calcium and 0.7 g of magnesium.[1] An adequate supply of these minerals to the developing fetus is dependent on placental transfer. In both human and rat pregnancy, net fetal accretion of calcium and magnesium rises exponentially in the last third of gestation[1-3] (Figure 1) and, in the case of calcium, this is paralleled by an increase in the unidirectional maternofetal flux of calcium across the placenta.[4] Current knowledge about the mechanism(s) which lead to this large increase in net transfer from mother to fetus of these two ions will be discussed later.

A. MATERNOFETAL GRADIENT

It has been known for many years that the concentration of total calcium in fetal blood is higher than that in maternal blood.[5] By using mid-gestation umbilical cord sampling of fetal blood in the human, it has been shown that this maternofetal gradient might be present as early as 20–26 weeks of gestation.[6] The demonstration that ionized calcium concentration in fetal blood is also greater than in maternal blood (Table 1) implies that the maternofetal gradient is not simply due to a difference in protein binding capacity between the two systems.

As in the case of calcium, the concentration of total and ultrafilterable magnesium in maternal plasma is lower than in fetal plasma in several species including the human (Table 2). However, since magnesium is principally an

Table 1 **Total and ionized calcium concentration in maternal and fetal blood (mmol/l ± SE)**

Species	Mother		Fetus		Ref.
	Total	Ionized	Total	Ionized	
Human	2.15 ± 0.01	1.12 ± 0.01	2.64 ± 0.02	1.41 ± 0.01	7
Monkey	2.05 ± 0.08	1.05 ± 0.06	2.10 ± 0.08	1.33 ± 0.05	8
Rat	2.23 ± 0.06	1.10 ± 0.03[a]	2.75 ± 0.02	1.42 ± 0.03[a]	9
Sheep	2.12 ± 0.05	0.99 ± 0.09	3.30 ± 0.05	1.59 ± 0.16	10
Pig	2.53 ± 0.02	0.79 ± 0.03	3.10 ± 0.04	1.12 ± 0.14	11

[a] Ionized calcium data provided in our laboratory by N. Robinson, 1988.

Table 2 **Total and ultrafilterable* concentration of magnesium in maternal and fetal plasma (mmol/l ± SE)**

Species	Mother	Fetus	Ref.
Human	0.63 ± 0.01	0.69 ± 0.01	7
Sheep	0.80 ± 0.17	0.93 ± 0.13	12
Rat	0.55 ± 0.02	0.84 ± 0.01	13
Rat*	0.36 ± 0.02	0.54 ± 0.04	13

intracellular ion, its plasma concentration might not reflect tissue magnesium levels. Total magnesium concentration within maternal erythrocytes is reportedly higher than in cord erythrocytes.[14] Nevertheless, the maternofetal gradient of plasma magnesium concentration is likely to be more important in the consideration of placental magnesium transfer (see below).

B. CONTROL OF FETAL PLASMA CALCIUM AND MAGNESIUM CONCENTRATIONS

Our understanding of the mechanisms by which plasma calcium and magnesium concentrations are maintained within a narrow range in postnatal life is poor. It is hardly suprising, therefore, that our understanding of these mechanisms during antenatal life is almost non-existent. The dependence of fetal plasma calcium and magnesium on the concentration of these cations in maternal plasma will be discussed later.

In the fetal rat over the last two to three days of gestation, there is an inverse relationship between plasma calcium and immunoreactive parathyroid hormone concentration which suggests a functional negative feedback mechanism between plasma calcium concentration and fetal parathyroid glands.[15] In sheep[16,17] and cows[18] acute fetal hypocalcemia induced by EDTA infusion leads to increased production of fetal parathyroid hormone. This is not so in the the fetal pig,[11] and conflicting results have been reported in the monkey,[19,20] which

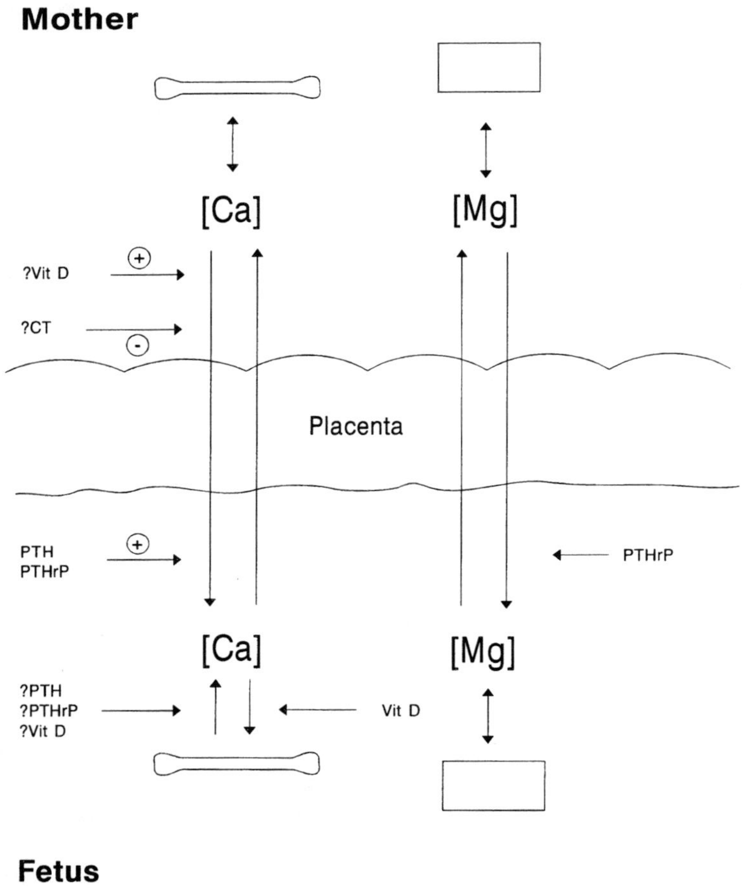

Figure 2 Cartoon showing bidirectional placental flux of calcium and magnesium. Both ions in body fluids are assumed to be freely interchangeable with their respective major pools, i.e. bone for calcium and cells for magnesium. The role of calciotropic hormones in maintaining fetal plasma calcium and magnesium concentrations are indicated by the horizontal arrows; where preceded by ?, the role is suggested but not proven. [Ca]: plasma calcium concentration; [Mg]: plasma magnesium concentration; Vit D: 1,25-dihydroxyvitamin D_3; CT: calcitonin; PTH: parathyroid hormone; PTHrP: parathyroid hormone-related peptide.

might reflect differences in methodology. In the sheep, pig and cow, acute fetal hypercalcemia leads to increased production of fetal calcitonin.[11,21]

Calciotropic hormones might modulate placental calcium and magnesium transport in addition to affecting the fetal reservoir of these cations (Figure 2). Therefore, interpretation of the data presented above is difficult and it is unclear whether the fetus can exert some autonomous control over its plasma concentration of calcium and magnesium.

II. ROUTES OF PLACENTAL TRANSFER

As in other transporting epithelia, it is likely that the placenta possesses both a paracellular and a transcellular route for ion transfer.[22,23] During human pregnancy, fetal capillary endothelium and the syncytiotrophoblast form two complete layers which separate maternal and fetal blood. The fetal capillary endothelium consists of apposing cells separated by a paracellular cleft with a width of about 16 nm[24] which is unlikely to hinder ion transfer. The syncytiotrophoblast, however, forms a true syncytium and is likely to be the major barrier in maternofetal ion exchange and the cell layer involved in transcellular transport processes.[25] Although the morphological correlate has not been adequately delineated, there is considerable physiological evidence for the presence of a paracellular route of diffusion across the syncytiotrophoblast. Recent studies indicate that paracellular routes in the human placenta may be formed by areas of syncytial denudation[26] or by tubular structures which are too small to be detectable under normal conditions.[27] By employing microelectrode techniques in the human placenta *in vitro*, Greenwood et al[28] found that the villous core was a mean of 4 mV negative with respect to the maternal side which would be sufficient to generate an electrochemical gradient to drive net transfer, of at least magnesium, to the fetus.[29] A maternofetal potential difference has been measured in several species[5] but the relationship of this to a transplacental potential difference is unclear.[30] Thus, the relative importance of the paracellular route for maternofetal calcium and magnesium transport is not known.

It is possible that calcium and magnesium could be transported across the trophoblast by endocytosis/exocytosis, i.e., ion uptake through formation of pinocytic vesicle at the maternal-facing placental plasma membrane, which then travels through the cytosol to the opposite pole of the trophoblast, followed by fusion to fetal-facing placental plasma membrane and release of ions into fetal plasma across fetal capillary endothelium. Calcium-containing vesicles within trophoblast cytosol have been identified by electron microscopy. Whereas endocytosis is the major mechanism for IgG transport across the human placenta,[5] there is no proof at the present time of placental calcium or magnesium transport by this mechanism.

In the following sections, discussion of placental calcium and magnesium transport will be confined mainly to transcelluar routes of transfer.

III. PLACENTAL TRANSPORT OF CALCIUM

A. ACTIVE TRANSPORT

Since the concentration of total, ultrafilterable and ionized calcium in fetal plasma is higher than in maternal plasma, calcium transport from mother to fetus across the placenta occurs against a concentration gradient and is likely to involve active transport mechanisms. This notion is supported by data from animal experiments in which the fetal circulation of the *in situ* placenta has been perfused and the maternofetal transfer of calcium measured. In the rat, Stulc and Stulcova[31] found that the maternofetal clearance of ^{45}Ca (37.4 ± 0.8 μl/min) across the *in situ* perfused placenta was much greater than that of inert, electrically neutral, hydrophilic molecules such as sucrose (2.6 ± 0.3 μl/min), which are believed to cross the placenta by passive diffusion. Using the same technique, Shaw et al[13] found that the unidirectional maternofetal clearance of ^{45}Ca across the *in situ* perfused rat placenta was significantly reduced when the perfusate temperature was lowered (Figure 3). Both studies[13,31] reported that maternofetal placental calcium transport was significantly reduced when the perfusate contained cyanide which suggests that placental calcium transport is dependent on placental metabolism (Figure 4). Twardock and Austin[32] perfused the fetal circulation of the guinea pig placenta *in situ* with plasma from an adult guinea pig and observed a significant rise in the total calcium concentration of perfusate effluent, even though the calcium concentration of inflowing perfusion fluid exceeded that of maternal plasma. Similarly, in the sheep *in situ* perfusion of the placenta with autologous fetal blood led to an increase in the ionized and total calcium concentrations of the perfusate with time,[33,34] despite the fact that the initial perfusate ionized and total calcium concentrations were higher than that of maternal plasma. In the human, Abramovich et al[35] perfused placental lobules *in vitro* and reported an increase in both ionized and total calcium concentrations in the fetal perfusate reservoir during recycling perfusion.

In conclusion, maternofetal transport of calcium across the placenta takes place against a concentration gradient, is temperature dependent, is inhibited by metabolic inhibitors such as cyanide, and is considerably greater than that of molecules which cross the placenta by simple diffusion. Taken together, these studies indicate that placental calcium transport involves active transfer processes via a transcellular route.

B. BIDIRECTIONAL TRANSPORT

Placental transport of calcium occurs both in the maternofetal and fetomaternal directions. The algebraic sum of these bidirectional fluxes determines the rate of net placental flux. It is clear that there is species variation in the rate of unidirectional maternofetal and fetomaternal fluxes. By using deconvolution analysis following the injection of calcium radioisotopes into maternal and fetal circulations of the sheep near term, Ramberg et al[36] reported unidirectional maternofetal flux of 212 mg/day/kg fetal weight and unidirectional fetomaternal flux of 12 mg/day/kg. This asymmetric flux is also found in the rat where net fetal calcium accretion between day 20 and 21 of gestation (term = 22 days) was reported to be 100 ± 4 nmol/min, a value similar to the unidirectional maternofetal flux (100 ± 7 nmol/min) estimated in nonanaesthetized animals on day 21 of gestation.[31] By contrast, in the rhesus monkey, unidirectional maternofetal and fetomaternal fluxes are similar (390 mg/day/kg fetal weight and 325 mg/day/kg, respectively).[36]

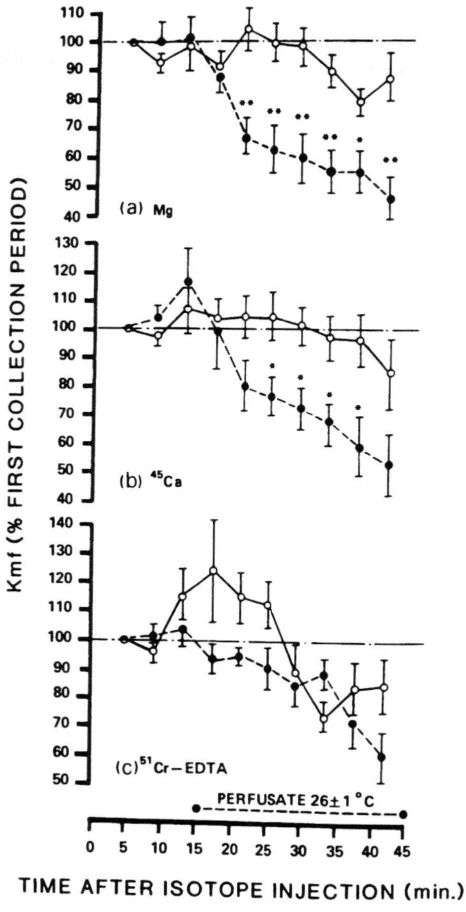

Figure 3 Effects of lowering temperature of perfusate fluid from 37 ± 1°C to 26 ± 1°C on maternofetal clearance (K_{mf}) of (a) Mg, (b) ^{45}Ca, and (c) ^{51}Cr-EDTA across *in situ* rat placentae perfused on fetal side. Mean K_{mf} values ± SEM are normalized to that for the first 4-min perfusate collection period (n = 6/group). Statistical significance between treated (•) and control (○) groups: $^{*}p < 0.05$; $^{**}p < 0.01$ (Student's t test, two-tailed). (From Shaw, A. J. et al., *Pediatr Res,* 27, 622, 1990. With permission.)

C. MECHANISMS OF TRANSPORT

There are at least three steps involved in transcellular placental calcium transport:

1. Influx of calcium ions from maternal plasma across microvillous placental membrane into the trophoblastic cytosol.
2. Translocation of calcium through the cytosol from microvillous to basal placental membrane.
3. Efflux of calcium ions from cytosol across the basal placental membrane and, hence, across fetal capillary endothelium into fetal plasma.

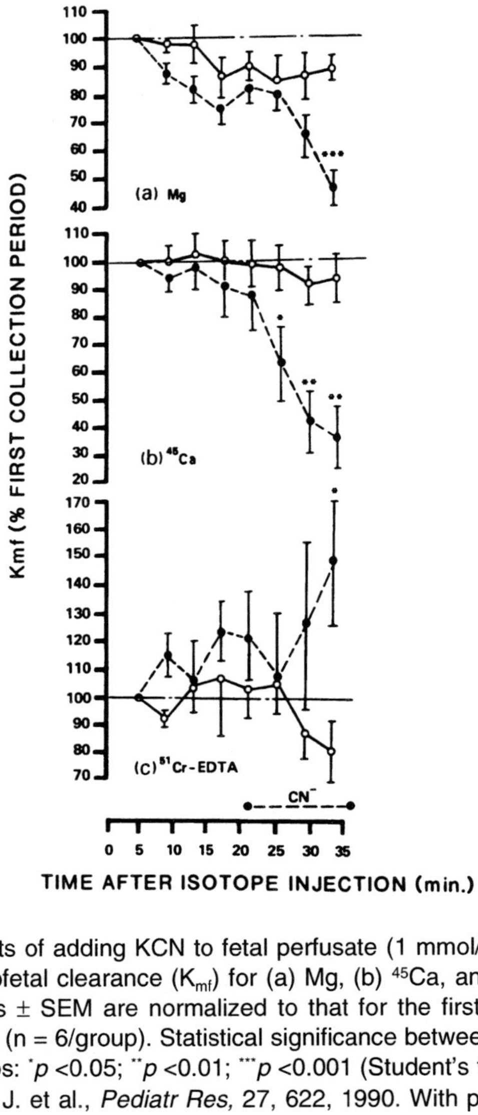

Figure 4 Effects of adding KCN to fetal perfusate (1 mmol/l final concentration) on maternofetal clearance (K_{mf}) for (a) Mg, (b) ^{45}Ca, and (c) ^{51}Cr-EDTA. Mean K_{mf} values ± SEM are normalized to that for the first 4-min perfusate collection period (n = 6/group). Statistical significance between treated (●) and control (○) groups: $^{*}p$ <0.05; $^{**}p$ <0.01; $^{***}p$ <0.001 (Student's t test, two-tailed). (From Shaw, A. J. et al., *Pediatr Res,* 27, 622, 1990. With permission.)

1. Influx

The precise mechanism by which calcium crosses the microvillous membrane and whether it crosses in the ionized or complexed form are unknown. In the guinea pig placenta, van Kreel and van Dijk[37] examined the effect of ^{45}Ca uptake from the maternal circulation on the indirectly measured microvillous transmembrane potential and suggested that calcium enters the trophoblast in the ionic form. The presence of saturable, carrier-mediated transport in both the microvillous and basal membranes of the guinea pig placenta has been demonstrated using the single circulation paired-tracer dilution method. In the guinea pig placenta, Michaelis-Menten kinetic constants on the fetal side were $Km = 1.8 \pm 0.7$ mM and $Vmax = 1.66 \pm 0.28$ μmol/min/g; on the maternal side, inter-animal variability was observed with a range of *Km* (0.18–1.15 mM) and *Vmax* (0.12–0.39 μmol/min/g) values.[38] Thus, in this species, the carrier-mediated transport system on the maternal side is of higher affinity and lower capacity compared to the fetal side and is likely to be almost fully saturated at physiological maternal plasma ionized calcium concentration. Stulc et al[39] perfused both the fetal and maternal circulations of the *in situ* rat placenta and also observed carrier-mediated transport characteristics on the maternal side characterized by kinetic constants similar to those reported in the guinea pig.

Sweiry et al[40] provided evidence of carrier-mediated calcium transport on both the maternal and fetal sides of the *in vitro* perfused human placenta using the single circulation paired-tracer dilution method. Recently, Kamath et al[41] have characterized the kinetic parameters of calcium uptake into human placental microvillous membrane vesicles. They described a two-site model with kinetic estimates of $Km = 18$ μM and 1.5 mM, $Vmax = 3.2$ and 47 nmol/min/mg, respectively.

These studies have contributed to our understanding of microvillous membrane calcium transport and the results indicate similarities with mediated entry mechanisms in intestine and kidney.[42-44] In these tissues, as in the placenta, *Km* values are approximately 1 mM. In the intestine, there is evidence that 1,25-dihydroxyvitamin D_3 (1,25$(OH)_2$D), the biologically active metabolite of vitamin D, increases *Vmax* of calcium influx, without altering the *Km*.[45] Factors other than the calcium concentration in the vicinity of the microvillous membrane which might control placental calcium influx are unknown.

2. Translocation

The free intracellular calcium concentration must be maintained in the micromolar range if cell viability is to be preserved. In other transporting epithelia, the resting intracellular calcium concentration is maintained at a value of 10^{-7} M, in comparison to the extracellular concentration of 10^{-3} M.[43] The concentration of free calcium in trophoblastic cytosol has not been measured but it is likely to be similar to that in other cells; in cultured syncytiotrophoblast the cytosolic concentration of free calcium is reported to be 4.5×10^{-8} M.[46] The large traffic of calcium ions across the trophoblast poses a threat to normal cellular function and, therefore, efficient mechanisms for intracellular calcium buffering and translocation from the maternal-facing cytosol to the fetal-facing cytosol must exist.

Intracellular organelles (nucleus, mitochondria, endoplasmic reticulum) and specific cytosol proteins known as calcium-binding proteins (CaBPs) may serve to maintain a low concentration of free calcium within the trophoblast by sequestrating excess calcium. Following perfusion with ^{45}Ca and differential centrifugation in the guinea pig placenta, van Kreel and van Dijk[37] demonstrated that most of the radiolabelled calcium was in the nucleus (48%) with lesser amounts bound to mitochondria (9%) and cytoplasmic proteins (28%). By complexing calcium with pyroantimonate into a granular form which could be detected using electron microscopy, Croley[47] observed granular deposits localized to endoplasmic reticulum and mitochondria and in vesicles within trophoblastic cytosol.

CaBPs have been identified in the placenta of the rat,[48] mouse,[49,50] and human.[51,52] Rat placental CaBP, comprised of 78 amino acids with a molecular weight of approximately 9000, is identical to the vitamin D-dependent CaBP in the intestine,[53] but does not seem to share the characteristic dependency of the latter on vitamin D. Maternal vitamin D deficiency in the rat results in a decrease in intestinal, but not placental, CaBP concentrations.[54] Verhaeghe et al[55] fed pregnant rats a low calcium-phosphorus diet and observed an increase in the concentration and free index (calculated as the $1,25(OH)_2D$ to vitamin D-binding protein molar ratio) of $1,25(OH)_2D$ in maternal plasma and a two-fold increase in duodenal CaBP; placental CaBP was unaffected. Similarly, injection of $1,25(OH)_2D$ in pregnant mice increases maternal intestinal CaBP but has no effect on placental CaBP concentrations.[49] Thyroparathyroidectomy in the pregnant rat resulted in reduced maternal plasma concentration of $1,25(OH)_2D$ accompanied by a 50% reduction in maternal duodenal CaBP concentration, whereas placental CaBP concentration was reduced by 20%, not significantly different from the control group.[56] Maternal treatment with $1,25(OH)_2D$ normalized the CaBP concentration in both tissues.[56] Thus, it appears that, unlike the intestine, placental CaBP is not controlled by maternal vitamin D status.

In the rat, a rapid increase in placental concentration of CaBP[48,57,58] and its mRNA[59,60] occurs during the period of increased rate of fetal growth and calcium accumulation in late gestation. Recently, Glazier et al[4] have reported a temporal association between placental CaBP mRNA expression and unidirectional maternofetal flux of calcium in rat placenta and have speculated that CaBP mRNA expression may be rate-limiting in placental calcium transport in this species (Figure 5). Whether similar changes occur in placental CaBP in other species is unknown and, therefore, the data in the rat should be interpretated with some caution. Indeed, the level of CaBP mRNA in human placenta does not change significantly from 14 to 32 weeks of gestation,[61] although what happens beyond this period when the fetus rapidly accumulates calcium has not been determined. Using standard molecular biology techniques, colleagues in our laboratory have tried to isolate CaBP mRNA from both human placental tissue at term and cytotrophoblast cell culture systems but with little success to date.[62]

Thus, the precise role of placental CaBPs remains speculative, but it is tempting to suggest that these proteins act as trophoblast cytosol buffers for calcium and/or facilitate translocation of calcium from maternal-facing cytosol to fetal-facing cytosol. There is some evidence to support the latter role. Tuan[52] demonstrated that uptake of calcium by human placental microsomal mem-

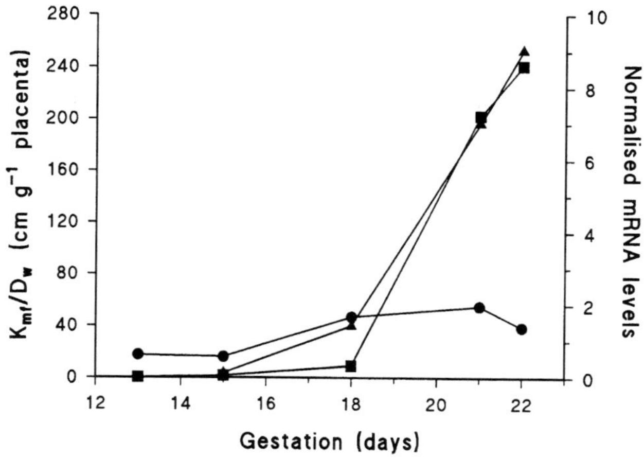

Figure 5 Relationship between maternofetal clearance/diffusion coefficient in water (K_{mf}/D_w) for ^{45}Ca (▲) and mRNA levels for calbindin$_{9K}$ (■) and Ca^{2+}-ATPase (●) in rat placenta during gestation. mRNA levels are shown as density values for calbindin$_{9K}$ or Ca^{2+}-ATPase mRNA normalized to β-actin mRNA density. (From Glazier, J. D. et al., *Am J Physiol,* 263, R930, 1992. With permission.)

brane vesicles was inhibited by preincubation with anti-human CaBP antibodies. CaBP enhances diffusion of calcium through an aqueous compartment[63] and Bronner et al[64] calculated that in the presence of CaBP the rate of cytosolic calcium diffusion in enterocytes is 60- to 70-fold higher than across cytosol of CaBP-depleted cells. CaBPs might act as an "intracellular shuttle" to facilitate the diffusion of calcium through trophoblastic cytosol, but more research in this area is required before their role can be defined.

3. Efflux

The final step in the maternofetal transtrophoblastic flux of calcium is efflux of calcium across the basal placental membrane. How this is acheived against a steep concentration gradient is not entirely clear. In other calcium transporting epithelia (e.g., intestine and kidney), there are at least two mechanisms involved in efflux of calcium ions out of the cytosol across the basolateral cell membrane:

1. Calcium- and magnesium-dependent adenosine triphosphatase (Ca^{2+}-ATPase).
2. Ca^{2+}/Na^{+} exchanger.

There is growing evidence that in the placenta a Ca^{2+}- ATPase may be responsible for the translocation of calcium from trophoblast cytosol to fetal extracellular fluid across the basal plasma membrane. Hydrolysis of ATP could provide the required energy to overcome the calcium concentration gradient. Ca^{2+}-ATPase has been identified in guinea pig[65] and human placental homogenate,[66] and in human placental plasma membrane vesicles.[67-69] It is clear that the Ca^{2+}-ATPases are a heterogenous group of enzymes, not all of which are involved in calcium transport,[70-72] which may explain the reported dissociation between Ca^{2+}-ATPase transport activity and Ca^{2+}-ATPase hydrolytic activity. Thus, Whitsett and Tsang[67] found Ca^{2+}-ATPase activity to be greater in placental mivrovillous membrane vesicles, whereas calcium uptake was greater in microsomal (containing microvillous, basal, and intracellular organelle membrane) vesicles. Similarly, Treinen and Kulkarni[68] reported both low and high affinity Ca^{2+}-ATPase activities in human placental microvillous membranes. Recently, however, calcium uptake into highly purified human placental basal plasma membrane vesicles has been described.[73,74] This high affinity uptake process, which is ATP- and magnesium-dependent and is stimulated by calmodulin (a calcium-binding protein), is likely to be mediated by one of the Ca^{2+}-ATPases. Using a monoclonal antibody raised against the human erythrocyte plasma membrane calcium pump and immunohistochemistry, Borke et al[75] have demonstrated greater specific staining in the human placental basal membrane and in the fetal facing membrane of the rat placenta. Furthermore, they also reported a 50% inhibition of calcium transport in basal plasma membrane vesicles from human placenta by polyclonal antibodies raised against the erythrocyte calcium pump.

The Ca^{2+}/Na^+ exchanger is a low affinity high capacity calcium transporter and is energized indirectly by means of the inwardly directed transmembrane Na^+ gradient maintained by $(Na^+ + K^+)$-ATPase. van Kreel and van Dijk,[37] who perfused both the maternal and fetal circulations of term guinea pig placenta with calcium-free media containing ^{22}Na and unlabelled sodium, found that addition of 2 mM Ca to the media after 30 minutes of equilibration did not affect the release of preloaded ^{22}Na from placental cells into the media. In the presence of a Ca^{2+}/Na^+ exchange system, this should have resulted in an increase in the rate of ^{22}Na release from placental cells. The presence of a Ca^{2+}/Na^+ exchanger has been sought in the *in situ* rat placenta by perfusing the fetal circulation and measuring the maternofetal flux of calcium in the presence or absence of sodium in the perfusate. In the absence of sodium, Stulc and Stulcova[31] found maternofetal flux of calcium to be unaltered whereas Shaw et al[76] reported a decrease. However, in the latter study, maternofetal calcium flux was unaffected during perfusion with sodium-containing perfusate to which ouabain (a potent $[Na^+ + K^+]$-ATPase inhibitor) had been added; by dissipation of the sodium gradient, an effect on calcium flux should have occurred in the presence of a Ca^{2+}/Na^+ exchanger. Whitsett and Tsang[67] noted that uptake of calcium by human placental microsomal vesicles was not inhibited by exclusion of sodium or addition of ouabain to the incubation medium; Lafond et al[74] reported similar findings when sodium was exluded from the incubation medium of highly purified basal membrane vesicles. Thus, present evidence indicates that the Ca^{2+}/Na^+ exchange system is unimportant in transplacental calcium movement.

D. CONTROL OF TRANSPORT

As mentioned previously, there is a rapid increase in the rate of net placental calcium transport (or the rate of calcium accretion in fetal skeleton) during the last third of pregnancy. The regulatory factors which trigger and maintain this increased maternofetal calcium flux are poorly understood. Supply of calcium to the fetus might be affected by maternal and fetal plasma calcium concentrations, maternal and fetal blood flow, and the capacity and efficiency of placental transport mechanisms.[5]

1. Blood Flow

In the rat, the maternofetal clearance of calcium per gram of placental tissue increases 70-fold in the last 6 days of gestation,[4] far more than could be expected by an increase in placental surface area or blood flow. In the human, it has been calculated that over the whole of pregnancy, 170 liters of maternal plasma would have to be cleared of its plasma calcium content to account for the total amount of calcium accumulated in a term fetus.[77] The estimated uterine blood flow rate in the human at term gestation is 500 ml/min[78] and, therefore, placental calcium transfer is likely to be limited by the area and the efficiency of placental transfer mechanisms rather than utero-placental blood flow.[77] However, Mughal et al[79] observed a marked decrease in the unidirectional maternofetal clearance of ^{45}Ca (but not of ^{51}Cr-EDTA, the reference diffusional marker) in growth-retarded rat fetuses induced by uterine artery ligation compared with that across placentae of control fetuses. The authors suggested that placental ischemia from chronic reduction in uteroplacental blood flow might result in depletion of the necessary energy supply for active placental calcium transport. Thus, reduced blood flow *per se* is unlikely to have a direct effect on placental calcium transport; however, alterations in placental metabolism as a result of placental ischemia might affect active transcellular calcium transport.

2. The Influence of Maternal Plasma Calcium Concentration

In the human, maternal and fetal serum ionized calcium concentrations are highly correlated[7,80] which suggests some dependence of fetal calcemia on maternal serum ionized calcium concentration. There is indirect evidence from human studies that the maternal plasma calcium concentration may influence placental calcium transport. Pregnant Chinese women, in whom plasma calcium concentrations were as low as 1.0 mM/l due to dietary deficiency, gave birth to infants with congenital rickets, whose plasma calcium concentrations were around 1.6 mM/l.[81] Transient neonatal hyperparathyroidism has been reported in infants born to mothers with hypocalcemia due to poorly treated hypoparathyroidism.[82] Conversely, transient neonatal hypoparathyrodism due to untreated maternal hyperparthyrodism has been reported in infants born to hypercalcemic mothers.[83] It is assumed that under these conditions net placental calcium transport is affected by the maternal plasma calcium concentration which leads to fetal hypo- or hypercalcemia which in turn leads to fetal/ neonatal hyper- or hypoparathyroidism, respectively. Thus, in the pregnant human, placental calcium transport appears to be dependent on the maternal plasma calcium concentration. Furthermore, placental calcium transport mechanisms do not appear to be saturated even when the mother is hypercalcemic.

Studies in animals have produced divergent findings. Whereas observations similar to those in the human have been made in the guinea pig,[84] pig,[11] and rabbit,[85] this is not the case in the sheep,[86] cow,[87] and rat.[88] In the latter species, a doubling of maternal plasma calcium concentration causes only a 30% increase in the maternofetal calcium flux across the *in situ* perfused rat placenta.[31] In this species it is likely that at physiological maternal plasma ionized calcium concentration the placental calcium transport system is almost maximally saturated. In contrast, studies using the *in situ* perfused guinea pig placenta have demonstrated that maternal hypercalcemia results in a significant increase in maternofetal calcium flux,[89] which may be due to submaximal saturation of placental transport mechanisms or to the existence of a quantitatively important and possibly unphysiologic paracellular leak pathway in this preparation.[90]

In the sheep,[91] pig[11], and cow[18] acute maternal hypocalcemia does not lead to any change in fetal plasma calcium concentrations. In contrast, acute maternal hypocalcemia in the rat[92] results in a fall in fetal calcium concentration and conflicting results have been reported in the monkey.[19,20] However, chronic maternal hypocalcemia in the rat, induced by maternal vitamin D deficiency and thyroparathyroidectomy, did not alter the maternofetal calcium flux across *in situ* perfused placentae, compared with sham operated animals.[93]

Thus, there is clearly a marked species variation in the fetal response to acute and chronic maternal hypocalcemia or hypercalcemia, which may be physiologic and/or methodologic.

3. Hormonal Regulation of Calcium Transport

The placenta has no nerve supply and, therefore, the factors that increase placental calcium transport capacity as gestation proceeds must be either genetically preprogrammed or hormones produced by the mother, fetus, or the placenta itself. Hormonal signals controlling placental calcium transfer might act on either the maternal or the fetal side of the placenta. Possible regulatory hormonal candidates include $1{,}25(OH)_2D$, parathyroid hormone (PTH), calcitonin or other peptides secreted by the parathyroid glands, e.g., parathyroid hormone-related protein (PTHrP). Infants born in summer months or who are small for gestational age are reported to have a lower bone mineral content at birth when compared to winter born infants or appropriate for gestational age infants, respectively.[94,95] The reason for this is not entirely clear but might be related to seasonal differences in maternal and fetal calciotropic hormones[95,96] and their speculative effect on placental transport of calcium and magnesium.

a. 1,25-Dihydroxyvitamin D_3 and 1-α-Hydroxycholecalciferol

There are clinical reports of maternal vitamin D deficiency causing neonatal rickets,[97] which are consistent with the suggestion that maternal vitamin D may be important for placental calcium transfer. The presence of cytosolic receptors for 1,25$(OH)_2$D in human and rat placentae[98] and of CaBP (which is a molecular marker of the action of 1,25$(OH)_2$D in tissues such as intestine) in human,[51,52] rat,[48,57] and mouse[49,50] placentae supports the idea that this hormonal form of vitamin D might indeed regulate placental calcium transport. Potential sources of 1,25$(OH)_2$D available to the fetus include that transferred from maternal plasma across the placenta,[99] that synthesized by the fetal kidneys,[100] and that synthesized by the placenta itself.[101] Since the fetal plasma concentration of 1,25$(OH)_2$D falls following fetal nephrectomy in sheep,[102] it appears that in this species the fetal kidney is a major source of this metabolite of vitamin D.

Although animal experiments have provided interesting data, the role of maternal vitamin D in placental calcium transfer remains unclear. Durand et al[103] found that the administration of 1,25$(OH)_2$D to pregnant guinea pigs increased the total calcium content of fetuses compared with control animals on days 40 and 50, but not day 66, of gestation. They also noted that at all three gestational periods studied, the maternoplacental, but not the maternofetal transfer of ^{45}Ca, was enhanced by the hormone. The same investigators examined the effect of maternal treatment with 1α-hydroxcholecalciferol (1αOHCC) on placental calcium transfer in sheep. A 12-day course of treatment led to a significant increase in the fetal content of calcium as well as maternofetal calcium transfer.[104] In this species, maternal prolactin injections also have been reported to increase placental calcium transfer and fetal calcium accretion;[105] the effect was enhanced by simultaneous administration of 1αOHCC and prevented by simultaneous administration of bromocriptine or calcitonin. As noted by the authors, prolactin stimulates the synthesis of 1,25$(OH)_2$D, and it may have been the latter hormone that affected placental calcium transfer in their studies.

If administration of vitamin D metabolites to the mother augments placental calcium transfer and fetal calcium accretion, then it would be expected that vitamin D deficiency might have the opposite effect. This is not so. In the rat, Brommage and De Luca[106] reported that maternal vitamin D deficiency had no effect on the maternofetal calcium gradient or the total fetal calcium content. Mughal et al[93] found that the total fetal calcium content and the maternofetal clearance of ^{45}Ca across the *in situ* perfused rat placenta were not affected by maternal vitamin D deficiency and thyroparathyroidectomy. Furthermore, maternal vitamin D deficiency in the rat leads to a significant reduction in intestinal, but not placental, CaBP concentration[54] and its mRNA.[62] Placental CaBP may have a role to play in placental calcium transfer (see above) and, therefore, these findings are consistent with the observations of Brommage and De Luca[106] and Mughal et al.[93] Thus, maternal vitamin D is unlikely to regulate placental calcium flux but its effect on placental CaBP requires further investigation.

The maternofetal total calcium gradient was found to be unaffected by fetal nephrectomy in the rat,[107] although injection of a pharmacologic dose of $1,25(OH)_2D$ (2 ng/fetus) into 19-day rat fetuses increased the fetal plasma total calcium concentration on day 21 of gestation compared with that in control fetuses.[108] Following fetal nephrectomy in the sheep, Ross et al[102] observed a marked decrease in the fetal plasma concentration of $1,25(OH)_2D$ with reversal of the usual maternofetal calcium gradient which was restored by infusion of $1,25(OH)_2D$ into the fetal circulation. Moore et al,[109] using a similar experimental protocol, obtained identical results, but they also observed a rise in fetal plasma phosphorus concentrations following fetal nephrectomy, which was prevented by daily intravenous injections of $1,25(OH)_2D$ into these fetuses. These authors suggested that the fall in fetal plasma ionized calcium concentrations occurs as a result of fetal hyperphosphatemia following fetal nephrectomy. However, these studies provide only indirect evidence for the effect of fetal $1,25(OH)_2D$ on placental calcium transport and do not rule out the possibility that $1,25(OH)_2D$ may influence fetal plasma calcium concentrations by a different mechanism, such as by causing fetal bone resorption.

In support of a direct permissive effect of $1,25(OH)_2D$ on placental calcium transfer is the observation of Robinson et al[110] that $1,25(OH)_2D$ stimulates maternofetal calcium flux across *in situ* perfused placentae of rat fetuses parathyroidectomized by gross decapitation[111] but not across placentae of intact fetuses. Because parathyroid hormone is an important trophic stimulator of $1,25(OH)_2D$ synthesis, it is likely that parathyroidectomized rat fetuses had a lower plasma $1,25(OH)_2D$ concentration than intact fetuses. Therefore, $1,25(OH)_2D$ probably stimulates placental calcium transfer only when fetal plasma concentrations of this hormone are low to start with.

b. Parathyroid Hormone and Fetal Parathyroid Hormone-like Peptide

In the rat, fetal parathyroidectomy by decapitation[110,112] or injection of antibovine PTH serum[113] leads to a fall in the plasma total and ionized calcium concentration, which is restored by PTH injection.[110,114] Similarly in the sheep, fetal thyroparathyroidectomy with thyroxine replacement results in a precipitate fall in fetal ionized calcium concentration within 24 hours[34,115-118] and a decreased rate of calcium accumulation in autologous fetal blood used to perfuse the *in situ* placenta.[34,116,118] However, addition of either rat PTH(1-34) or bovine PTH(1-84) to the perfusate did not increase the rate of calcium accumulation.[116] As previously mentioned, Robinson et al[110] observed a significant fall in maternofetal clearance of ^{45}Ca across the *in situ* perfused placentae of decapitated rat fetuses. The clearance was increased (but not returned to normal levels) by subcutaneous administration of bovine PTH(1-84) or $1,25(OH)_2D$ to the fetuses. However, there was no effect on the maternofetal clearance of ^{45}Ca

when the perfusate medium contained a synthetic rat parathyroid hormone amide (r[$Nle^{8,21}$,Tyr^{34}]PTH(1-34)) in decapitated fetuses.[110] Thus, the stimulatory effect of PTH(1-84) on maternofetal ^{45}Ca clearance might be due to its direct action on the placental "calcium pump" or alternatively may act indirectly by increasing fetal 1,25$(OH)_2$D production which, in turn, might stimulate placental calcium transfer (see above). Perfusion of the fetal circulation of intact[110,119] and decapitated[110] rat fetuses with perfusion media containing forskolin (an activator of adenylate cyclase[120]) stimulated maternofetal clearance of ^{45}Ca. Since r[$Nle^{8,21}$,Tyr^{34}]PTH(1-34) can also increase placental cyclic AMP release,[119] the role of placental adenylate cyclase in placental calcium transfer is not clear.

There is evidence that hormones other than PTH acting on the fetal side of the placenta might be candidates as enhancers of placental calcium transport. Using a two-site sensitive immunoradiometric assay specific for intact human PTH, two recent reports have demonstrated a lower concentration of this hormone in human umbilical plasma compared to maternal plasma.[121,122] In contrast, in the pig,[11] rat,[123] and human,[121,124] the concentration of bioactive PTH in fetal plasma, measured using the cytochemical assay of Chambers et al,[125] was greater than in maternal plasma. Furthermore, extracts of fetal parathyroid gland contain a substance which could not be neutralized by excess antiserum to PTH.[116,126,127] Based on these observations it appears that fetal parathyroid glands produce a second substance which acts functionally but not immunochemically as PTH. Addition of fetal parathyroid gland extract to autologous fetal blood used to perfuse placentae of thyroparathyroidectomized sheep fetuses *in situ* led to an increase in the rate of calcium accumulation in the fetal blood reservoir.[116,117,126,128] Thus, it appears that a PTH-like peptide produced in fetal parathyroid glands might be capable of stimulating maternofetal calcium transport. Using the same experimental model, Rodda et al[116] demonstrated that addition of fetal parathyroid gland extract or PTHrP (isolated from a human cancer cell line), but not of bovine PTH(1-84) or rat PTH(1-34), resulted in further increases in the total plasma calcium concentration of blood used to perfuse placentae of thyroparathyroidectomized sheep fetuses. These authors suggested that the PTH-like peptide produced by fetal parathyroid glands is probably an oncofetal peptide, similar to the hypercalcemic PTHrP, a PTH-like hypercalcemic peptide expressed by some solid tumors.[129] Recent work in the sheep suggests that mid-molecule fragments of PTHrP might be responsible for the stimulation of placental calcium transport.[117,128,130] However, this is not true in the rat[119] which may reflect methodological or species differences.

c. Calcitonin

In sheep, Barlet et al[131] demonstrated that chronic maternal calcitonin deficiency, achieved by maternal thyroidectomy on day 30 of gestation with subsequent thyroxine replacement, led to a significant increase in the total fetal calcium content and fetal accumulation of ^{45}Ca during the latter half of pregnancy. Calcitonin replacement in these ewes led to normalization of both the fetal calcium content and the maternofetal transfer of ^{45}Ca. Thus, in this species, maternal calcitonin appears to decrease the maternofetal calcium flux and its role, if any, in other species remains to be investigated.

IV. PLACENTAL TRANSPORT OF MAGNESIUM

A. ACTIVE TRANSPORT

Near the end of pregnancy the plasma concentrations of total and ultrafilterable magnesium in the fetus exceed that of the mother[13,132] (Table 2). Therefore, similar to placental calcium transport, maternofetal magnesium transfer occurs against a concentration gradient and is likely to involve active transport mechanisms. Evidence for active transcellular transport was provided by the studies of Shaw et al,[13] who reported that the unidirectional maternofetal clearance of magnesium across *in situ* perfused rat placentae was considerably higher (26.7 $\mu l\ min^{-1}\ g^{-1}$) than that of the extracellular marker, ^{51}Cr-EDTA (3.2 $\mu l\ min^{-1}\ g^{-1}$). These studies also demonstrated that the maternofetal clearance of magnesium was significantly reduced by lowering of perfusate temperature (Figure 3) or addition of cyanide to fetal perfusate (Figure 4).

B. BIDIRECTIONAL TRANSPORT

In human pregnancy, the fetal magnesium content increases rapidly after the fifth month of gestation, with net fluxes of some 5 mg/day.[1,133] Similar to calcium, magnesium transfer is likely to occur in both directions across the placenta. Using ^{28}Mg as a tracer, Care et al[134] demonstrated bidirectional placental magnesium fluxes in the near-term sheep placenta with an approximate maternofetal flux of 0.042 mg/hr/kg fetal weight and fetomaternal flux of 0.012 mg/hr/kg fetal weight. Fetomaternal flux has not been reported in other species.

C. MECHANISMS OF TRANSPORT

As for calcium, at least three steps are required for transplacental magnesium transfer.[29] Influx of magnesium across the maternal facing placental membrane might be a passive process depending on the intracellular free magnesium concentration and transmembrane potential difference. Membrane proteins mediating this influx have not been identified but might be magnesium channels or facilitated type carriers. The mechanism by which magnesium is translocated across the placental cytoplasm is unknown but might be by simple diffusion either in the free or bound state. Magnesium efflux across the fetal facing placental membrane is likely to be an energy-dependent process.

It is clear that the mechanisms involved in the transport of magnesium across the placenta are separate from those of calcium. Using the *in situ* perfused rat placenta, Shaw et al[76] demonstrated that manoeuvers which rapidly reduced maternofetal clearance of magnesium either had no effect on the clearance of calcium or did so with a different time course. This is in agreement with the lack of effect of dietary-induced hypomagnesemia on calcium clearance in the same species.[135]

In the rat, Gunther et al[136] reported that the transfer of ^{28}Mg from mother to fetus was reduced if amiloride or furosemide was administered to the pregnant animal and suggested the presence of a Na^+/Mg^{2+} exchanger and Mg^{2+}/HCO_3^- cotransporter based on the effects of these drugs in other tissues.[137] The driving force for these transporters could be provided by electrochemical gradients for sodium (generated by $Na^+ + K^+$ ATPase) and bicarbonate (generated by the dissociation of carbonic acid). The studies of Shaw et al[76] provided further evidence for the presence of a Na^+/Mg^{2+} exchanger. They found that the maternofetal clearance of magnesium across the *in situ* perfused rat placenta was reduced when the perfusate was either sodium free or contained ouabain, both of which would reduce the electrochemical gradient for sodium. However, the presence of acetazolamide (a carbonic anhydrase inhibitor) in the perfusate had no effect on the clearance of magnesium and, therefore, the role of a Mg^{2+}/HCO_3^- cotransporter remains speculative.[76]

D. CONTROL OF TRANSPORT

There is little information on the regulation of placental magnesium transport. Like calcium, net placental magnesium transfer must increase rapidly during the latter third of gestation to meet the demands of the growing fetus. Until a clearer picture of placental transport mechanisms emerges, the control of magnesium transport will remain poorly understood.

1. The Influence of Maternal Plasma Magnesium Concentration

In the rat there appears to be some dependence of placental magnesium transport on maternal plasma magnesium concentration. Using the dually perfused *in situ* rat placenta, Shaw et al[13] demonstrated that reduction of magnesium in the maternal perfusate by 90% reduced the appearance of magnesium in the fetal perfusate by 87%. Maternal hypomagnesemia in the rat produced by dietary deficiency leads to reduced fetal plasma and carcass magnesium concentrations and a loss of the fetomaternal magnesium gradient.[132,138-142] On the other hand, Garel et al[143] observed that an intraperitoneal injection of magnesium into pregnant rats resulted in nearly a three-fold increase in plasma concentration of magnesium in the mother, but only a slight increase in fetal plasma magnesium concentration. In sheep, neither maternal hypomagnesemia induced by 1α-hydroxyvitamin D_3 injections nor maternal hypermagnesemia induced by intravenous infusion of magnesium chloride altered the fetal plasma magnesium concentration.[144] Thus, in this species, placental magnesium transport appears to be saturated even when the mother is hypomagnesemic. In addition, any diffusional component of transfer must be very small in relation to the carrier mediated component.

In humans, treatment of pre-eclampsia with intravenous infusion of magnesium sulfate results in hypermagnesemia in both the mother and the fetus or neonate.[145-147] From this observation, it appears that the placental magnesium transporting mechanism is not saturated at physiological plasma magnesium concentrations or that magnesium is also transported by diffusion in this species. These inter-species variations in fetal response to maternal hypo- and hypermagnesemia might be related to differences in placental structure.

2. Hormonal Regulation of Magnesium Transport

Using *in situ* placental perfusion in sheep, it was reported that magnesium transport was increased by fetal parathyroid gland extract and mid-molecule fragments of human PTHrP (hPTHrP[67-86]amide and hPTHrP[75-86]amide), whereas the amino terminal (hPTHrP(1-34)) had no effect.[117,130] In the rat, however, Shaw et al[119] did not find a stimulatory effect on placental magnesium transport of hPTHrP(75-86)amide, hPTHrP(1-34) or r[Nle^{8,21},Tyr^{34}]PTH(1-34). These contradictory findings may again represent methodological or species differences.

V. DIABETIC PREGNANCY

Disturbances in fetal growth are commonly encountered in diabetic pregnancy in both animals and humans. In the rat, fetal growth retardation is a consistent finding irrespective of whether the disease is induced with streptozotocin or is genetic.[148-151] This contrasts with the situation in the diabetic pregnant woman, whose offspring tend to be overgrown at term,[152] although in the first trimester of such pregnancies the fetus might be growth retarded and more similar to the condition in rats.[153] A wide variety of qualitative and quantitative changes[154-158] has been reported in placentae of diabetic pregnancies but there is little data on measurement of placental nutrient flux in such pregnancies.

By using photon absortiometry, Mimouni et al[159] demonstrated that the bone mineral content in infants of diabetic mothers is reduced by about 10% when compared to normal infants. Fetal histomorphometric analysis and ashing studies have demonstrated similar findings in the offspring of both genetically diabetic rats and those rendered diabetic by streptozotocin.[148-151] These data suggest that net fetal accretion of bone minerals and fetal bone mineralization in diabetic pregnancy is reduced. Whereas the increased incidence of hypocalcemia and hypomagnesemia reported in infants of diabetic mothers might be due to disturbances in calciotropic hormone responsiveness,[160-165] reduced fetal bone mineral accretion is likely to be associated with lower net placental transport of bone minerals.

We have investigated this in more detail by measuring the maternofetal flux of calcium and magnesium across the *in situ* perfused rat placenta of the streptozotocin-induced diabetic rat. Our results indicate that in the presence of untreated maternal diabetes the maternofetal flux of both calcium and magnesium is reduced by between 20 and 40% to values significantly lower than in a control group of animals (Figure 6). These alterations in placental transport

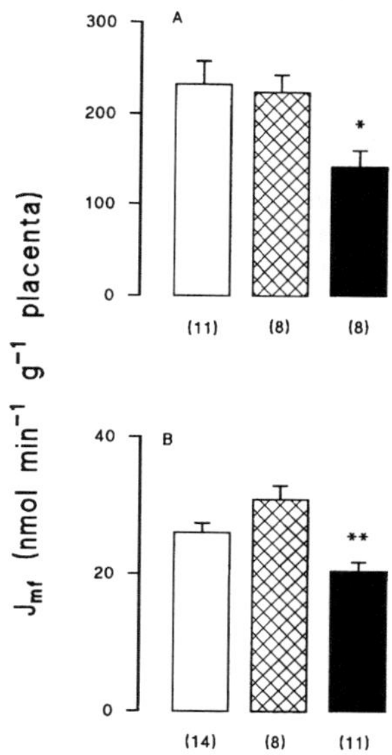

Figure 6 Unidirectional maternofetal flux (J_{mf}) per g placenta of (A) Ca and (B) Mg in control (C, open bars), insulin-treated diabetic (D_I, hatched bars), and untreated diabetic (D_O, solid bars) rats on day 21 of gestation. Number of animals is shown in parentheses. Data are means ± SEM. $^*p < 0.05$ v C and D_I; $^{**}p < 0.05$ v C and < 0.01 v D_I by Duncan's multiple range tests. (From Husain, S. M. et al., *Pediatr Res,* 35, 376, 1994. With permission.)

were prevented by control of the diabetes with insulin (Figure 6) and were unaccompanied by changes in passive placental permeability.[166] Net fetal accretion of calcium and magnesium was reduced in the untreated diabetic group of animals and largely normalized in the insulin-treated group. In agreement with previous reports, we found that the fetuses of the untreated diabetic group weighed significantly less than the other groups.[166] When allowance was made for this, we found that the calcium content, but not the magnesium content, remained significantly lower in the untreated diabetic group of animals (Figure 7). Thus, the net flux of calcium across the diabetic rat placenta is reduced to a greater degree than that expected from growth retardation alone. This is different from the situation in which growth retardation is induced in the rat by reducing uterine blood flow where net calcium flux is appropriate for fetal size.[79]

The pathophysiology of these disturbances in placental transport of calcium and magnesium during diabetic rat pregnancy is not fully understood and it is likely that a combination of factors lead to altered placental transport and fetal accretion of calcium and magnesium. Hypomagnesemia[167] and hypermagnesuria[168] are common in diabetes mellitus and we have reported an additive effect of pregnancy in increasing renal magnesium excretion.[169] However, our observations

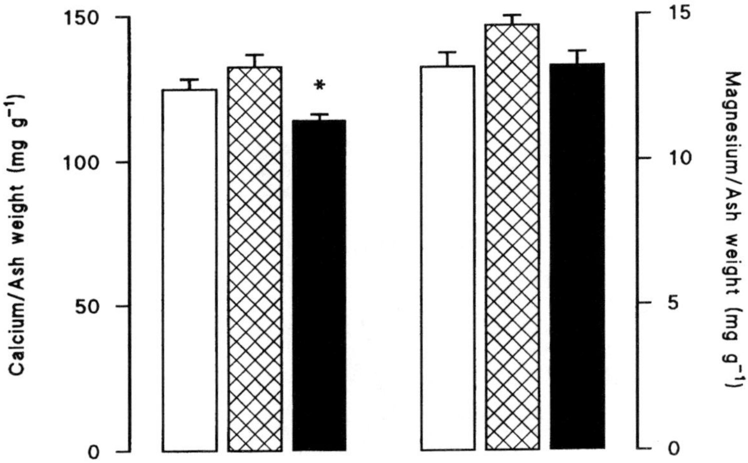

Figure 7 Fetal Ca and Mg content normalized to ash weight in control (C, open bars, n = 14), insulin-treated diabetic (D_I, hatched bars, n = 8), and untreated diabetic (D_O, solid bars, n = 11) rats on day 21 of gestation. Four fetuses were used from each animal; n is the number of pregnancies. Data are means ± SEM. **p* <0.05 v C and <0.01 v D_I by Duncan's multiple range tests. (From Husain, S. M. et al., *Pediatr Res,* 35, 376, 1994. With permission.)

in untreated diabetic rat pregnancy do not support reduced cation availability as a cause of the reduced maternofetal flux of these cations.[166] The finding by Verhaeghe et al[55] that CaBP is reduced in the placentae of diabetic rats is of interest since it is speculated that the expression of this protein might be rate limiting in placental calcium transport.[4] Our data demonstrate that the expression of CaBP mRNA in placentae of diabetic rats is reduced by a factor of 12–15 whereas cyclophilin mRNA (used as a control for non-specific differences) was only reduced by a factor of 1–2 (Figure 8). This is of considerable importance since this is the second situation where this laboratory has found a relationship between placental flux of calcium and its CaBP mRNA content.

Our understanding of fetal calcium and magnesium metabolism and placental transport of these cations is rudimentary with many unanswered questions: how does the fetus maintain a higher concentration of plasma calcium and magnesium compared to its mother? what are the precise mechanisms of placental transfer and how are they controlled? how do these factors interact to satisfy fetal demand? what "switches on" the increased rate of net calcium and magnesium flux across the placenta during the latter part of gestation? These are only a few issues which need to be addressed before a comprehensive understanding of fetal growth and metabolism can be acheived. Such knowledge might enable us to optimize fetal growth which might be a determinant of disease in adult life.[170,171]

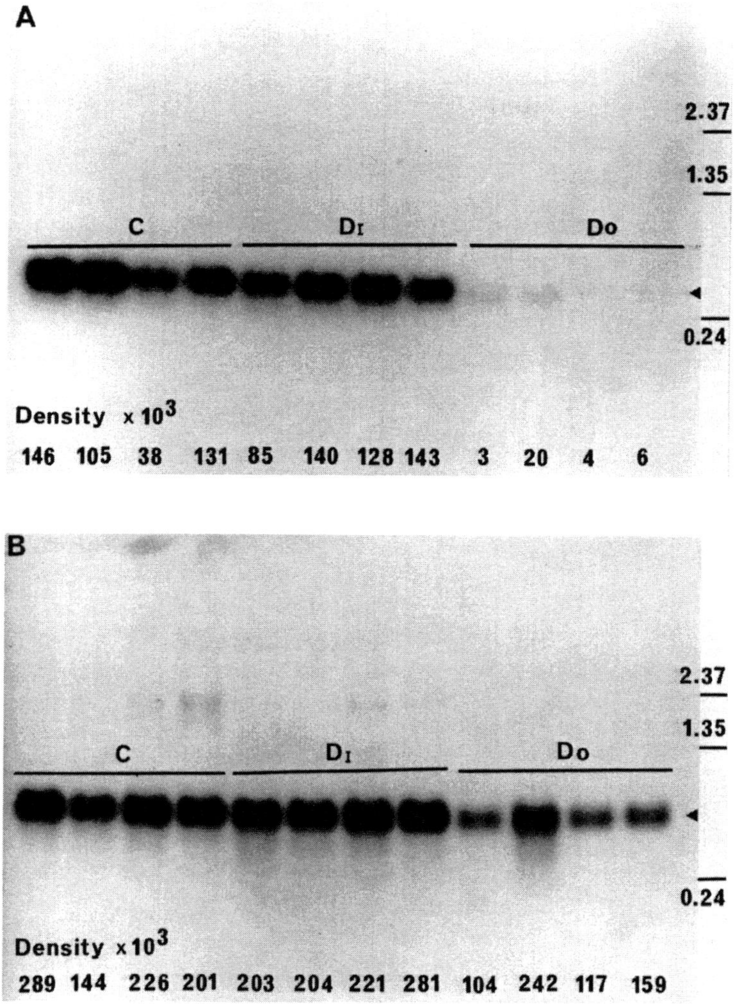

Figure 8 Northern hybridization of rat placental RNA with ^{32}P-cDNA probes. Single Northern blot probed with calbindin$_{9K}$ (A) and cyclophilin (B). Equal amounts of RNA (15μg) were loaded in lanes 1–4 control (C), 5–8 insulin-treated diabetic (D_I), and 9–12 untreated diabetic (D_O). Position of RNA size markers (in kb) is shown on right. Arrows indicate major RNA species at 0.44 kb (A) and 0.69 kb (B). Transcript density peak area values (arbitary units) are given beneath each lane. (From Husain, S. M. et al., *Pediatr Res,* 35, 376, 1994. With permission.)

REFERENCES

1. Widdowson, E. M., Changes in body composition during growth, in *Scientific Foundations of Paediatrics*, Davis, J. A., Dobbing, J., Eds., William Heinemann Medical Books Ltd, London, 1981, 330.
2. Comar, C. L., Radiocalcium studies in pregnancy, *Ann NY Acad Sci*, 64, 281, 1956.
3. Romeu, A., Alemany, M., Arola, L., Net transfer of essential metals from mother to fetus in the second half of pregnancy, *Biol Neonate*, 49, 204, 1986.
4. Glazier, J. D., Atkinson, D. E., Thornburg, K. L., Sharpe, P. T., Edwards, D., Boyd, R. D. H., Sibley, C. P., Gestational changes in Ca^{2+} transport across rat placenta and mRNA for calbindin_{9K} and Ca^{2+}-ATPase, *Am J Physiol*, 263, R930, 1992.
5. Sibley, C. P., Boyd, R. D. H., Control of transfer across the mature placenta, in *Oxford Reviews of Reproductive Biology*, Vol 10, Clarke, J. R., Ed., Oxford University Press, Oxford, 1988, 382.
6. Forestier, F., Daffos, F., Rainaut, M., Bruneau, M., Trivin, F., Blood chemistry of normal human fetuses at midtrimester of pregnancy, *Pediatr Res*, 21, 579, 1987.
7. Pitkin, R. M., Cruikshank, D. P., Schauberger, C. W., Reynolds, W. A., Williams, G. A., Hargis. G. K., Fetal calciotropic hormones and neonatal calcium homeostasis, *Pediatrics*, 66, 77, 1980.
8. Northrop, G., Misenhimer, H. R., Becker, F. O., Failure of parathyroid hormone to cross the nonhuman primate placenta, *Am J Obstet Gynecol*, 129, 449, 1977.
9. Chalon, S., Garel, J-M., Plasma calcium control in the rat fetus. I. Influence of maternal hormones, *Biol Neonate*, 48, 313, 1985.
10. Care, A. D., Ross, R., Pickard, D. W., Weatherley, A. J., Robinson, J. S., The role of the kidney in ovine foetal calcium and phosphate homeostasis, *Adv Physiol Sci*, 20, 45, 1981.
11. Care, A. D., Ross, R., Pickard, D. W., Weatherley, A. J., Garel, J-M., Manning, R. M., Allgrove, J., Papapoulos, S., O'Riordan, J. L. H., Calcium homeostasis in the fetal pig, *J Dev Physiol*, 4, 85, 1982.
12. Mellor, D. J., Matheson, I. C., Variations in the distribution of calcium, magnesium and inorganic phosphorus within chronically catheterized sheep conceptuses during the last eight weeks of pregnancy, *Q J Exp Physiol*, 62, 55, 1977.
13. Shaw, A. J., Mughal, M. Z., Mohammed, T., Maresh, M. J. A., Sibley, C. P., Evidence for active maternofetal transfer of magnesium across the *in situ* perfused rat placenta, *Pediatr Res*, 27, 622, 1990.
14. Paunier, L., Girardin, E., Brioschi, P. A., Beguin, F., Maternal-fetal relationship of extra- and intracellular magnesium and potassium concentration, in *Magnesium in Cellular Processes and Medicine*, Altura, B. M., Durlach, J., Seelig, M. S., Eds., Karger, Basel, 1987, 151.
15. Thomas, M. L., Anast, C. S., Forte, L. R., Regulation of calcium homeostasis in the fetal and neonatal rat, *Am J Physiol*, 240, 367, 1981.
16. Smith, F. G., Alexander, D. P., Buckle, R. M., Britton, H. G., Nixon, D. A., Parathyroid hormone in foetal and adult sheep: the effect of hypocalcemia, *J Endocrinol*, 53, 339, 1972.

17. Care, A. D., Pickard, D. W., Garel, J-M., Barlet, J. P., Tomlinson, S., O'Riordan, J. L. H., Autonomy of calcium metabolism in the sheep foetus, *Horm Metab Res*, 7, 103, 1975.
18. Wadsworth, J. C., Kronfeld, D. S., Ramberg, C. F., Parathyrin and calcium homeostasis in the fetus, *Biol Neonate*, 41, 101, 1982.
19. Fleischman, A. R., Lerman, S., Oakes, G. K., Epstein, M. F., Chez, R. A., Mintz, D. H., Perinatal primate parathyroid hormone metabolism, *Biol Neonate*, 27, 40, 1975.
20. Pitkin, R. M., Reynolds, W. A., Williams, G. A., Kawahara, W., Bauman, A. F., Hargis, G. K., Maternal and foetal parathyroid hormone responsiveness in pregnant primates, *J Clin Endocrinol Metab*, 51, 1044, 1980.
21. Littledike, E. T., Arnaud, C. D., Whipp, S. C., Calcitonin secretion in ovine, porcine, and bovine fetuses, *Proc Soc Exp Biol Med*, 139, 428, 1972.
22. Stulc, J., Extracellular transport pathways in the haemochorial placenta, *Placenta*, 10, 113, 1989.
23. Sibley, C. P., Boyd, R. D. H., Mechanisms of transfer across the human placenta, in *Fetal and Neonatal Physiology*, Polin, R. A., Fox, W. W., Eds., WB Saunders, Philadelphia, 1992, 62.
24. Leach, L., Firth, J. A., Fine structure of the paracellular junctions of terminal villous capillaries in the perfused human placenta, *Cell Tissue Res*, 268, 447, 1992.
25. Shennan, D. B., Boyd, C. A. R., Ion transport by the placenta: a review of membrane transport systems, *Biochim Biophys Acta*, 906, 437, 1987.
26. Edwards, D., Jones, C. J. P., Sibley, C. P., Nelson, D. M., Paracellular permeability pathways in the human placenta: a quantitative and morphological study of maternal-fetal transfer of horseradish peroxidase, *Placenta*, 14, 63, 1993.
27. Kertschanska, S., Kaufmann, P., Morphological evidence for the existence of transtrophoblastic channels in human placental villi, *Placenta*, 13, A.33, 1992.
28. Greenwood, S. L., Boyd, R. D. H., Sibley, C. P., Transtrophoblast and microvillous membrane potential difference in mature intermediate human placental villi, *Am J Physiol*, 265, C460, 1993.
29. Husain, S. M., Sibley, C. P., Magnesium and pregnancy, *Miner Electrolyte Metab*, 19, 296, 1993.
30. McNaughton, T., Power, G. G., The maternal-fetal electrical potential difference: new findings and a perspective, *Placenta*, 12, 185, 1991.
31. Stulc, J., Stulcova, B., Transport of calcium by the placenta of the rat, *J Physiol (Lond)*, 371, 1, 1986.
32. Twardock, A. R., Austin, M. K., Calcium transfer in perfused guinea pig placenta, *Am J Physiol*, 219, 540, 1970.
33. Weatherley, J., Ross, R., Pickard, D. W., Care, A. D., The transfer of calcium during perfusion of the placenta in intact and thyroparathyroidectomized sheep, *Placenta*, 4, 271, 1983.
34. Care, A. D., Caple, I. W., Abbas, S. K., Pickard, D. W., The effect of fetal thyroparathyroidectomy on the transport of calcium across the ovine placenta to the fetus, *Placenta*, 7, 417, 1986.
35. Abramovich, D. R., Dacke, C. G., Elcock, C., Page, K. R., Calcium transport across the isolated dually perfused human placental lobule. *J Physiol (Lond)*, 382, 397, 1987.

36. Ramberg, C. F., Delivoria-Papadopoulos, M., Crandall, E. D., Kornfield, D. S., Kinetic analysis of calcium transport across the placenta, *J Appl Physiol*, 35, 682, 1973.
37. van Kreel, B. K., van Dijk, J. P., Mechanisms involved in the transfer of calcium across the isolated guinea pig placenta, *J Develop Physiol*, 5, 155, 1983.
38. Sweiry, J. H., Yudilevich, D. L., Asymmetric calcium influx and efflux at maternal and fetal sides of the guinea pig placenta: kinetics and specificity, *J Physiol (Lond)*, 355, 295, 1984.
39. Stulc, J., Stulcova, B., Svihovec, J., Transport of calcium across the dually perfused placenta of the rat, *J Physiol (Lond)*, 420, 295, 1990.
40. Sweiry, J. H., Page, K. R., Dacke, C. G., Abramovich, D. R., Yudilevich, D. L., Evidence of saturable uptake mechanisms at the maternal and fetal sides of the perfused human placenta by rapid paired-tracer dilution: studies with calcium and choline, *J Devel Physiol*, 8, 435, 1986.
41. Kamath, S. G., Kelley, L. K., Friedman, A. F., Smith, C. H., Transport and binding in calcium uptake by microvillous membrane of human placenta, *Am J Physiol*, 262, C789, 1992.
42. Ghijsen, W. E. J. M., Ganguli, G., Strange, P. G., Murer, H., Calcium uptake into rat small intestinal brush border membrane vesicles: characterization of transmembrane calcium transport at short initial incubation times, *Cell Calcium*, 8, 157, 1987.
43. van Os, C. H., Transcellular calcium transport in intestinal and renal epithelial cells, *Biochim Bioiphys Acta*, 906, 195, 1987.
44. Bronner, F., Transcellular calcium transport, in *Intracellular Calcium Regulation*, Bronner, F., Ed., Liss, New York, 1990, 415.
45. Laing, C. T., Barnes, J., Balakir, R. A., Sacktor, B., Characterization of 1,25 dihydroxyvitamin D_3-dependent calcium uptake in isolated chick duodenal cells, *J Mem Biol*, 90, 145, 1986.
46. Illsley, N. P., Sellers, M. C., Jacobs, M. M., Endothelin-stimulated elevation of intracellular calcium in cultured human syncytiotrophoblast, in *Proceedings of the Society for Gynecologic Investigation 38th Annual Meeting*, San Antonio, 1991, 222.
47. Croley, T. E., The intracellular localization of calcium within the mature human placental barrier, *Am J Obstet Gynecol*, 117, 926, 1973.
48. Bruns, M. E. H., Fausto, A., Avioli, L. V., Placental calcium binding protein in rats. Apparent identity with vitamin D-dependent calcium binding protein from the rat intestine, *J Biol Chem*, 253, 3186, 1978.
49. Bruns, M. E. H., Wallshein, V., Bruns, D. E., Regulation of calcum-binding protein in mouse placenta and intestine, *Am J Physiol*, 242, E47, 1982.
50. Tuan, R. S., Cavanaugh, S. T., Identification and characterization of a calcium-binding protein in the mouse chorioallantoic placenta, *Biochem J*, 233, 41, 1985.
51. Tuan, R. S., Identification and characterization of a calcium-binding protein from human placenta, *Placenta*, 3, 145, 1982.
52. Tuan, R. S., Calcium-binding protein of the human placenta: characterization, immunohistochemical localization and functional involvement in Ca^{2+} transport, *Biochem J*, 227, 317, 1985.

53. MacManus, J. P., Watson, D. C., Yaguchi, M., The purification and complete amino acid sequence of the 9000-M_r Ca^{2+}-binding protein from rat placenta, *Biochem J*, 235, 585, 1986.
54. Marche, P., Delorme, A., Cuisnier-Gleizes, P., Intestinal and placental calcium-binding proteins in vitamin D-deprived or -supplemented rats, *Life Sci*, 23, 2555, 1978.
55. Verhaeghe, J., Thomasset, M., Brehier, A., van Assche, F. A., Bouillon, R., $1,25(OH)_2D_3$ and Ca-binding protein in fetal rats: relationship to the maternal vitamin D status, *Am J Physiol*, 254, E505, 1988.
56. Garel, J-M., Delorme, A. C., Marche, P., Nguyen, T. M., Garabedian, M., Vitamin D_3 metabolite injections to thyroparathyroidectomized pregnant rats: effects on calcium-binding proteins of maternal duodenum and of fetoplacental unit, *Endocrinology*, 109, 284, 1981.
57. Delorme, A. C., Marche, P., Garel, J. M., Vitamin D-dependent calcium-binding protein. Changes during gestation, prenatal and postnatal development in rats, *J Dev Physiol*, 1, 181, 1979.
58. Mathieu, C. L., Burnett, S. H., Mills, S. E., Overpeck, J. G., Bruns, D. E., Bruns, M. E., Gestational changes in calbindin-D_{9k} in rat uterus, yolk sac, and placenta: implications for maternal-fetal calcium transport and uterine muscle function, *Proc Natl Acad Sci USA*, 86, 3433, 1989.
59. de Maintenant, F., Perret, C., Brun, P., Thomasset, M., Expression of cholecalcin (9000 MW cholecalciferol-induced CaBP) gene in rat duodenum and placenta during gestation, in *Vitamin D. A Chemical, Biochemical and Clinical Update*, Norman, A. W., Schaefer, K., Grigolett, H. G., Herrath, D. V., Eds., de Gruyter, Berlin, 1985, 355.
60. Krisinger, J., Dann, J. L., Jeung, E. B., Leung, P. C. K., Calbindin-D_{9k} gene expression during pregnancy and lactation in the rat, *Mol Cell Endocrinol*, 88, 119, 1992.
61. Brun, P., Dupret, J. M., Thomasset, M., Mathieu, H., Vitamin D-dependent calcium-binding proteins in human fetuses: comparative distribution of 9K CaBP mRNA and 28K CaBP during development, *Pediatr Res*, 21, 362, 1987.
62. Glazier, J. D., personal communication, 1993.
63. Feher, J. J., Facilitated calcium diffusion by intestinal calcium-binding protein, *Am J Physiol*, 244, C303, 1983.
64. Bronner, F., Pansu, D., Stein, W. D., An analysis of intestinal calcium transport across the rat intestine, *Am J Physiol*, 250, G561, 1986.
65. Shami, Y., Radde, I. C., Calcium-stimulated ATPase of guinea pig placenta, *Biochim Biophys Acta*, 249, 345, 1971.
66. Miller, R. K., Berndt, W. O., Evidence for Mg^{2+}-dependent $(Na^{+}+K^{+})$-activated ATPase and Ca^{2+}-ATPase in the human placenta, *Proc Soc Exp Biol Med*, 143, 118, 1973.
67. Whitsett, J. A., Tsang, R. C., Calcium uptake and binding by membrane fractions of the human placenta, *Pediatr Res*, 14, 769, 1980.
68. Treinen, K. A., Kulkarni, A. P., High-affinity, calcium- stimulated ATPase in brush border membranes of the human term placenta, *Placenta*, 7, 365, 1986.
69. Tuan, R. S., Kushner, T., Calcium-activated ATPase of the human placenta: identification, characterization, and functional involvement in calcium transport, *Placenta*, 8, 53, 1987.

70. Kelley, L. K., Smith, C. H., Use of GTP to distinguish calcium transporting ATPase activity from other calcium dependent nucleotide phosphatases in human placental basal plasma membrane, *Biochem Biophys Res Comm*, 148, 126, 1987.
71. Kelley, L. K., Borke, J. L., Verma, A. K., Kumar, R., Penniston, J. T., Smith, C. H., The calcium-transporting ATPase and the calcium- or magnesium-dependent nucleotide phosphatase activities of human placental trophoblast basal plasma membrane are separate enzyme activities, *J Biol Chem*, 265, 5433, 1990.
72. Wuytack, F., Raeymaekers, L., The Ca^{2+}-transport ATPases from the plasma membrane, *J Bioenergetics Biomembranes*, 24, 285, 1992.
73. Fisher, G. H., Kelly, L. K., Smith, C. H., ATP-dependent calcium transport across basal plasma membranes of human placental trophoblast, *Am J Physiol*, 252, C38, 1987.
74. Lafond, J., Leclerc, M., Brunette, M. G., Charaterization of calcium transport by basal plasma membranes from human placental syncytiotrophoblast, *J Cell Physiol*, 148, 17, 1991.
75. Borke, J. L., Caride, A., Verma, A. K., Kelley, L. K., Smith, C. H., Penniston, J. T., Kumar, R., Calcium pump epitopes in placental trophoblast basal plasma membranes, *Am J Physiol*, 257, C341, 1989.
76. Shaw, A. J., Mughal, M. Z., Maresh, M. J. A., Sibley, C. P., Sodium-dependent magnesium transport across in-situ perfused rat placenta, *Am J Physiol*, 261, R369, 1991.
77. Canning, J. F., Boyd, R. D. H., Mineral and water exchange between mother and fetus, in *Fetal Physiology and Medicine*, Beard, R. W., Nathanielsz, P. W., Eds., Marcel Dekker Inc., New York, 1984, 481.
78. Blechner, J. N., Stenger, V. G., Prystowsky, H., Uterine blood flow in woman at term, *Am J Obstet Gynecol*, 120, 633, 1974.
79. Mughal, M. Z., Ross, R., Tsang, R. C., Clearance of calcium across *in-situ* perfused placentas of intrauterine growth retarded rat fetuses, *Pediatr Res*, 25, 420, 1989.
80. Schauberger, C. W., Pitkin, R. M., Maternal-perinatal calcium relationships, *Obstet Gynecol*, 53, 74, 1979.
81. Maxwell, J. P., Further studies of adult rickets (osteomalacia) and fetal rickets, *Proc R Soc Med*, 28, 265, 1934.
82. Landing, B. H., Kamoshita, S., Congenital hyperparathyroidism secondary to maternal hypoparathyroidism, *J Pediatr*, 77, 842, 1970.
83. Jacobson, B. B., Tersley, E., Lund, B., Sorensen, O. H., Neonatal hypocalcemia associated with maternal hyperparathyroidism, *Arch Dis Child*, 53, 308, 1978.
84. Durand, D., Dalle, M., Barlet, J-P., Plasma calcium homeostasis in the guinea pig during the perinatal period, *Biol Neonate*, 42, 120, 1982.
85. Graham, R. W., Porter, G. P., Fetal maternal plasma calcium relationships in the rabbit, *Q J Exp Physiol*, 56, 160, 1971.
86. Bawden, J. W., Wolkoff, A. S., Fetal blood calcium responses to maternal calcium infusion in sheep, *Am J Obstet Gynecol*, 99, 55, 1967.
87. Barlet, J-P., Davicco, M-J., LeFaivre, J., Carillo, B. J., Fetal blood calcium response to maternal hypercalcemia induced in the cow by calcium infusion or by Solanium glaucophyllum ingestion, *Horm Metab Res*, 11, 57, 1979.

88. Chalon, S., Garel, J-M., Plasma calcium control in the rat fetus. III. Influence of alterations in maternal plasma calcium on fetal plasma calcium levels, *Biol Neonate*, 48, 329, 1985.
89. Derewlany, L. O., Radda, I. C., Transplacental ^{45}Ca and ^{32}P flux in the guinea pig: effect of maternal hypercalcemia and hypocalcemia, *Can J Physiol Pharmacol*, 63, 1577, 1985.
90. Hedley, R., Bradbury, M. W. B., Transport of polar non- electrolytes across the intact and perfused guinea pig placenta, *Placenta*, 1, 277, 1980.
91. Care, A. D., Calcium homeostasis in the fetus, *J Dev Physiol,* 2, 85, 1980.
92. Garel, J-M., Dumont, C., Barlet, J-P., Care, A., D., Fetal-maternal calcium relationships in rat and sheep, *J Physiol (Paris)*, 64, 387, 1972.
93. Mughal, M. Z., Robinson, N. R., Sibley, C. P., Materno-fetal calcium (Ca) transfer across the in situ perfused rat placenta: relation to maternal 1,25-Dihydroxyvitamin D (1,25D) administration, *Pediatr Res*, 433A, 1987.
94. Namgung, R., Mimouni, F., Campaigne, B. N., Ho, M. L., Tsang, R. C., Low bone mineral content in summer-born compared with winter-born infants, *J Pediatr Gastroenterol Nutr*, 15, 285, 1992.
95. Namgung, R., Tsang, R. C., Specker, B. L., Sierra, R. I., Ho, M. L., Reduced serum osteocalcin and 1,25-dihydroxyvitamin D concentrations and low bone mineral content in small for gestational age infants: evidence of decreased bone formation rates, *J Pediatr*, 122, 269, 1993.
96. Verity, C. M., Burman, D., Beadle, P. C., Holton, J. B., Morris, A., Seasonal changes in perinatal vitamin D metabolism: maternal and cord blood biochemistry in normal pregnancies, *Arch Dis Child*, 56, 943, 1981.
97. Park, W., Paust, H., Kaufmann, H. J., Offermann, G., Osteomalacia of the mother — rickets of the newborn, *Eur J Pediatr*, 147, 292, 1987.
98. Pike, J. W., Gooze, L. L., Haussler, M. R., Biochemical evidence for 1,25-dihydroxyvitamin D receptor macromolecules in parathyroid, pancreatic, pituitary and placental tissues, *Life Sci*, 26, 407, 1980.
99. Ron, M., Levitz, M., Chuba, J., Dancis, J., Transfer of 25-hydroxyvitamin D_3 and 1,25-dihdroxynitamin D_3 across the perfused human placenta, *Am J Obstet Gynecol*, 148, 370, 1984.
100. Fenton, E., Britton, H. G., 25-Hydroxycholecalciferol 1α-hydroxylase activity in the kidney of the foetal, neonatal and adult guinea pig, *Biol Neonate*, 37, 256, 1980.
101. Whitsett, J. A., Ho, M., Tsang, R. C., Norman, E. J., Adams, K. G., Synthesis of 1,25-dihydroxyvitamin D_3 by human placenta in vitro, *J Clin Endocrinol Metab*, 53, 486, 1981.
102. Ross, R., Care, A. D., Robinson, J. S., Pickard, D. W., Weatherley, A. J., Perinatal 1,25-dihydroxycholecalciferol in the sheep and its role in the maintenance of the transplacental calcium gradient, *J Endocrinol*, 87, 17, 1980
103. Durand, D., Barlet, J-P., Braithwaite, G. D., The influence of 1,25-dihydroxycalciferol on the mineral content of foetal guinea pigs, *Reprod Nutr Dev*, 23, 235, 1983.
104. Durand, D., Braithwaite, G. D., Barlet, J-P., The effect of 1α-hydroxycalciferol on the placental transfer of calcium and phosphate in sheep, *Br J Nutr*, 49, 475, 1983.
105. Barlet, J-P., Prolactin and calcium metabolism in pregnant ewes, *J Endocrinol*, 107, 171, 1985.

106. Brommage, R., De Luca, H. F., Placental transport of calcium and phosphorus is not regulated by vitamin D, *Am J Physiol*, 246, F526, 1984.
107. Chalon, S., Garel, J-M., Plasma calcium control in the rat fetus. II. Influence of fetal hormones, *Biol Neonate*, 48, 323, 1985.
108. Chalon, S., Garel, J-M., 1,25-dihydroxyvitamin D_3 injections into rat fetuses: effects on fetal plasma calcium, plasma phosphate and mineral content, *Reprod Nutr Dev*, 23, 567, 1983.
109. Moore, E. S., Langman, C. B., Favus, M. J., Coe, F. L., Role of fetal 1,25-dihydroxyvitamin D production in intrauterine phosphorus and calcium metabolism, *Pediatr Res*, 19, 566, 1985.
110. Robinson, N. R., Sibley, C. P., Mughal, M. Z., Boyd, R. D. H., Fetal control of calcium transport across the rat placenta, *Pediatr Res*, 26, 109, 1989.
111. Jost, A., Experiences de decapitation de l'embryon de Lapin, *C R Acad Sci Serie B*, 225, 322, 1947.
112. Pic, P., Maniey, J., Jost, A., Facteur endocriniens reglant la calcemie et de la phosphoremie du foetus du rat, *C R Soc Biol*, 159, 1274, 1965.
113. Garel, J-M., Effet de l'injection d'un serum 'antiparathormone' chez le foetus de rat, *C R Acad Sci Serie D*, 271, 2364, 1970.
114. Garel, J-M., Pic, J., Jost, A., Action de la parathormone chez le foetus de rat, *Ann Endocrinol*, 32, 253, 1971.
115. Care, A. D., Ross, R., Fetal calcium homeostasis, *J Dev Physiol*, 6, 59, 1984.
116. Rodda, C. P., Kubota, M., Heath, J. A., Moseley, J. M., Care, A. D., Caple, I. W., Martin, T. J., Evidence for a novel parathyroid hormone-related protein in fetal lamb parathyroid glands and sheep placenta: comparisons with a similar protein implicated in humoral hypercalcaemia of malignancy, *J Endocrinol*, 117, 261, 1988.
117. Barri. M., Abbas, S. K., Pickard, D. W., Hammonds, R. G., Wood, W. I., Caple, I. W., Martin, T. J., Care, A. D., Fetal magnesium homeostasis in the sheep, *Exp Physiol*, 75, 681, 1990.
118. MacIsaac, R. J., Heath, J. A., Rodda, C. P., Moseley, J. M., Care, A. D., Martin, T. J., Caple, I. W., Role of the fetal parathyroid glands and parathyroid hormone-related protein in the regulation of placental transport of calcium, magnesium and inorganic phosphate, *Reprod Fertil Dev*, 3, 447, 1991.
119. Shaw, A. J., Mughal, M. Z., Maresh, M. J. A., Sibley, C. P., Effects of two synthetic parathyroid hormone-related protein fragments on maternofetal transfer of calcium and magnesium and release of cyclic AMP by the in-situ perfused rat placenta, *J Endocrinol*, 129, 399, 1991.
120. Seamon, K. B., Daly, J. W., Forskolin: a unique diterpene activator of cyclic AMP-generating systems, *J Cyclic Nucleotide Protein Phosphor Res*, 7, 201, 1981.
121. Rubin, L. P., Posillico, J. T., Anast, C. S., Brown, E. M., Circulating levels of biologically active and immunoreactive intact parathyroid hormone in human newborns, *Pediatr Res*, 29, 201, 1991.
122. Saggese, G., Baroncelli, G. I., Bertelloni, S., Cipolloni, C., Intact parathyroid hormone levels during pregnancy, in healthy term neonates and in hypocalcemic preterm infants, *Acta Paediatr Scand*, 80, 36, 1991.
123. Bourdeau, A., Manganella, G., Thil-Trubert, C-L., Sachs, C., Cournot, G., Bioactive parathyroid hormone in pregnant rats and fetuses, *Am J Physiol*, 258, E549, 1990.

124. Allgrove, J., Adami, S., Manning, R. M., O'Riordan, J. L. H., Cytochemical bioassay of parathyroid hormone in maternal and cord blood, *Arch Dis Child*, 60, 110, 1987.
125. Chambers, D. J., Dunham, J., Zanelli, J. M., Parsons, J. A., Bitensky, L., Chayen, J., A sensitive bioassay of parathyroid hormone in plasma, *Clin Endocrinol*, 9, 375, 1978.
126. Abbas, S. K., Caple, I. W., Care, A. D., Loveridge, N., Martin, T. J., Pickard, D. W., Rodda, C., The role of the parathyroid glands in fetal calcium homeostasis in the sheep, *J Physiol (Lond)*, 386, 27P, 1987.
127. Loveridge, N., Caple, I. W., Rodda, C., Martin, T. J., Care, A. D., Further evidence for a parathyroid hormone-related protein in fetal parathyroid glands of sheep, *Q J Exp Physiol*, 73, 781, 1988.
128. Abbas, S. K., Pickard, D. W., Rodda, C. P., Heath, J. A., Hammonds, R. G., Wood, W. I., Caple, I. W., Martin, T. J., Care, A. D., Stimulation of ovine placental calcium transport by purified natural and recombinant parathyroid hormone-related protein (PTHrP) preparations, *Q J Exp Physiol*, 74, 549, 1989.
129. Suva, L. J., Winslow, G. A., Wettenhall, R. E. H., Hammonds, R. G., Mossley, J. M., Diefenbach-Jagger, H., Rodda, C. P., Kemp, B. E., Rodrigeuz, H., Chen, E. Y., Hudson, P. J., Martin, T. J., Wood, W. I., A parathyroid hormone-related protein implicated in malignant hypercalcemia. Cloning and expression, *Science*, 237, 893, 1987.
130. Care, A. D., Abbas, S. K., Pickard, D. W., Barri, M., Drinkhill, M., Findlay, J. B. C., White, I. R., Caple, I. W., Stimulation of ovine placental transport of calcium and magnesium by mid-molecule fragments of human parathyroid hormone-related protein, *Exp Physiol*, 75, 605, 1990.
131. Barlet, J-P., Calcitonin may modulate placental transfer in ewes, *J Endocrinol*, 104, 17, 1985.
132. Dancis, J., Springer, D., Cohlan, S. Q., Fetal homeostasis in maternal malnutrition. II. Magnesium deprivation, *Pediatr Res*, 5, 131, 1971.
133. Givens, M. H., Machy, I. G., The chemical composition of the human fetus, *J Biol Chem*, 102, 7, 1933.
134. Care, A. D., Pickard, D. W., Weatherley, A., The measurement of transplacental magnesium fluxes in the sheep, *Res Vet Sci*, 27, 121, 1979.
135. Cruz, M. L., Mimouni, F., Tsang, R. C., Hammond, G., Effect of chronic maternal dietary magnesium deficiency on placental calcium transport, *J Am Coll Nutr*, 11, 87, 1992.
136. Gunther, T., Vormann, J., Hollriegl, V., Effects of amiloride and furosemide on ^{28}Mg transport into fetuses and maternal tissues of rats, *Magnesium Bull*, 10, 34, 1988.
137. Beyenbach, K. W., Transport of magnesium across biological membranes, *Magnesium Trace Elem*, 9, 233, 1990.
138. Cohlan, S. Q., Jansen, V., Dancis, J., Piomelli, S., Microcytic anemia with erythroblastosis in offspring of magnesium-deprived rats, *Blood*, 36, 500, 1970.
139. Cosens, G., Diamond, I., Theriault, L. L., Hurley, L. S., Magnesium deficiency anemia in the rat fetus, *Pediatr Res*, 11, 758, 1977.
140. Vormann, J., Forster, R., Gunther, T., Foetal and maternal magnesium metabolism: effect of maternal magnesium deficiency and isoproterenol, *J Clin Chem Clin Biochem*, 21, 765, 1983.

141. Vormann, J., Gunther, T., Development of fetal mineral and trace element metabolism in rats with normal as well as magnesium- and zinc-deficient diets, *Biol Trace Elem Res*, 9, 37, 1986.
142. Giavini, E, Broccia, M. L., Prati, M., Roversi, G. D., Vismara, C., The minimum content of magnesium in the diet for a normal pregnancy in the rat, *Nutr Res*, 8, 509, 1988.
143. Garel, J-M., Barlet, J-P., Calcitonin in the mother, fetus and newborn, *Ann Biol Anim Biochim Biophys*, 18, 53, 1978.
144. Barlet, J-P., Davicco, M. J., Moncoup, M., LeFaivere, J., Foetal plasma magnesium levels during maternal hypo- or hypermagnesemia in ewes, *Br J Nutr*, 42, 559, 1979.
145. Cruikshank, D. P., Pitkin, R. M., Reynolds, W. A., Williams, G. A., Hargis, G. K., Effects of magnesium sulfate treatment on perinatal calcium metabolism I. Maternal and fetal responses, *Am J Obstet Gynecol*, 134, 243, 1979.
146. Donovan, E. F., Tsang, R. C., Steichen, J. J., Strub, R. J., Chen, I-W., Chen, M., Neonatal hypermagnesemia: effect on parathyroid hormone and calcium homeostasis, *J Pediatr*, 96, 305, 1980.
147. Rasch, D. K., Huber, P. A., Richardson, C. J., L'Hommedieu, C. S., Nelson, T. E., Reddi, R., Neurobehavioral effects of neonatal hypermagnesemia, *J Pediatr*, 100, 272, 1982.
148. Uriu-Hare, J. Y., Stern, J. S., Reaven, G. M., Keen, C. L., The effect of maternal diabetes on trace element status and fetal development in the rat, *Diabetes,* 34, 1031, 1985.
149. Verhaeghe, J., Bouillon, R., Nyomba, B. L., Lissens, W., van Assche, F. A., Vitamin D and bone mineral homeostasis during pregnancy in the diabetic BB rat, *Endocrinology*, 118, 1019, 1986.
150. Demignon, J., Bonneton-Rebut, C., Effects of experimental diabetes on the vitamin D metabolism of pregnant rats and their fetuses, *Calcif Tissue Int*, 42, 127, 1988.
151. Verhaeghe, J., Bouillon, R., Lissens, W., Visser, W. J., van Assche, F. A., Diabetes and low Ca-P diet have opposite effects on adult and fetal bone mineral metabolism, *Am J Physiol*, 254, E496, 1988.
152. Sosenko, I. R., Kitzmiller, J. L., Loo, S. W., Blix, P., Rubenstein, A. H., Gabbay, K.H., The infant of the diabetic mother. Correlation of increased cord C-peptide levels with macrosomia and hypoglycemia, *N Engl J Med*, 301, 859, 1979.
153. Pedersen, J. F., Mølsted, L., Early growth retardation in diabetic pregnancy, *Brit Med J*, 1, 18, 1979.
154. Diamant, Y. Z., Metzger, B. E., Freinkel, N., Shafrir, E., Placental lipid and glycogen content in human and experimental diabetes mellitus, *Am J Obstet Gynecol*, 144, 5, 1982.
155. Eriksson, U. J., Jansson, L., Diabetes in pregnancy: decreased placental blood flow and disturbed fetal development in the rat, *Pediatr Res*, 18, 735, 1984.
156. Teasdale, F., Histomorphometry of the human placenta in class C diabetes mellitus, *Placenta*, 6, 69, 1985.
157. Boyd, P. A., Scott, A., Keeling, J. A., Quantitative structural studies on placentas from pregnancies complicated by diabetes mellitus, *Br J Obstet Gynaecol*, 93, 31, 1986.

158. Robinson, J. R., Canavan, J. P., el Haj, A. J., Goldspink, D. F., Maternal diabetes in rats I. Effects on placental growth and protein turnover, *Diabetes*, 37, 1665, 1988.
159. Mimouni, F., Steichen, J. J., Tsang, R. C., Hertzberg, V., Miodovnik, M., Decreased bone mineral content in infants of diabetic mothers, *Am J Perinatol*, 5, 339, 1988.
160. Tsang, R. C., Kleinman, L. I., Sutherland, J. M., Light, I. J., Hypocalcemia in infants of diabetic mothers. Studies in calcium phosphorus and magnesium metabolism and parathormone responsiveness, *J Pediatr*, 80, 384, 1972.
161. Bergman, L., Kjellmer, I., Selstam, U., Calcitonin and parathyroid hormone - relation to early neonatal hypocalcemia in infants of diabetic mothers, *Biol Neonate*, 24, 151, 1974.
162. Tsang, R. C., Strub, R., Brown, D. R., Steichen, J., Hartman, C., Chen, I-W., Hypomagnesemia in infants of diabetic mothers: perinatal studies, *J Pediatr*, 89, 115, 1976.
163. Salle, B., David, F., Glorieux, F., Delvin, E. E., Loius, J. J., Troncy, G., Hypocalcemia in infants of diabetic mothers. Studies in circulating calciotropic hormone concentrations, *Acta Pediatr Scand*, 71, 573, 1982.
164. Mimouni, F., Tsang, R. C., Hertzberg, V. S., Miodovnik, M., Polycythemia, hypomagnesemia, and hypocalcemia in infants of diabetic mothers, *Am J Dis Child*, 140, 798, 1986.
165. Mimouni, F., Loughead, J. L., Tsang, R. C., Khoury, J., Postnatal surge in serum calcitonin concentrations: no contribution to neonatal hypocalcemia in infants of diabetic mothers, *Pediatr Res*, 28, 493, 1990.
166. Husain, S. M., Birdsey, T. J., Glazier, J. D., Mughal, Z. M., Garland, H. O., Sibley, C. P., Maternofetal calcium flux and calbindin$_{9k}$ expression in the diabetic rat placenta, *Pediatr Res*, 35, 376, 1994.
167. Jackson, C. E., Meier, D. W., Routine serum magnesium analysis, *Ann Int Med*, 69, 743, 1968.
168. Garland, H. O., New experimental data on the relationship between diabetes mellitus and magnesium, *Magnesium Res*, 5, 193, 1992.
169. Birdsey, T. J., Husain, S. M., Garland, H. O., Sibley, C. P., Mughal, M. Z., The effect of diabetes mellitus on renal magnesium handling in the pregnant rat, in *Procceedings of the XXXII International Congress of Physiological Sciences*, Glasgow, 1993, 109.
170. Barker, D. J. P., Osmond, C., Golding, J., Kuh, D., Wadsworth, M. E. J., Growth in utero, blood pressure in childhood and adult life, and mortality from cardiovascular disease, *Br Med J*, 298, 564, 1989.
171. Barker, D. J. P., Bull, A. R., Osmond, C., Simmonds, S. J., Fetal and placental size and risk of hypertension in adult life, *Br Med J*, 301, 259, 1990.

Chapter 3

Calcium and Magnesium Metabolism in Preterm Infants

Winston W. K. Koo and Francis Mimouni

CONTENTS

0-8493-4613-4/95/$0.00+$.50

I. PHYSIOLOGIC REGULATION OF CALCIUM AND MAGNESIUM HOMEOSTASIS

In preterm infants, limited information indicates that the mechanisms involved in calcium (Ca) and magnesium (Mg) homeostasis are maintained in accordance with physiologic principles similar to term infants (see Chapter 4) and older children (see Chapter 6). Developmentally, calciotropic hormones are found early in human gestation and appear to be functional at birth in preterm infants. Immunoreactive parathyroid hormone containing cells are detected in parathyroid glands of human fetuses greater than ten weeks gestation.[1] Immunoradiometric assays for intact parathyroid hormone (PTH) show similar serum concentrations between term and preterm infants but are lower than normal adult values.[2,3] However, cytochemical bioassays show elevated bioactivity of PTH in both term and preterm infants compared to normal adults and mothers, and umbilical venous PTH bioactivity is highly correlated with the magnitude of the transplacental (difference in plasma Ca concentration between umbilical vein and maternal blood) Ca gradient.[2] Furthermore, preterm infants are shown to have an appropriate rise in both bioactivity and immunoreactive intact PTH concentrations in association with postnatal decline in plasma total and ionized Ca concentrations.[2,3] Calcitonin content in crude tissue preparation of thyroid is larger than that of the adult thyroid.[4] Serum calcitonin concentrations are high at birth compared to paired maternal serum calcitonin concentrations. They further increase as early as the first few hours after birth and may reach values five- to tenfold higher than adult values.[5-7] With Ca infusion, serum calcitonin concentrations increase and are opposite to the changes in serum PTH concentrations.[6] This is consistent with their physiologic role in maintaining Ca homeostasis and modulating the effect of other calciotropic hormones. Serum calcitonin concentrations gradually decrease during infancy although the mean serum calcitonin concentrations in preterm infants several months after birth still may be twice the adult values.[5,7] Whether this is related to "catch up" bone mineralization after the initial illness in preterm infants is not known. It is now generally accepted that vitamin D metabolism in preterm infants is normal and appropriate for the functional needs of these infants. It is intimately related to both mineral metabolism and skeletal mineralization for the developing preterm infant (for details, see Section III).

Regulation of Mg homeostasis in the preterm infant appears to be less dependent on calciotropic hormones, although physiologically Mg is important for PTH secretion[8] and affects target organ responsiveness to PTH.[9] Mg also is an essential co-factor in vitamin D metabolism, and hypomagnesemia is associated with low serum 1,25 dihydroxyvitamin D concentrations.[10] As with older pediatric and adult subjects, intestinal absorption and regulation of renal excretion appear to be the major regulators of Mg homeostasis and also are important in regulation of Ca homeostasis.[11-14] In addition, Ca and Mg may each affect the regulation of the other at the intestinal and renal levels.[15]

There are also circumstances unique to the preterm infant that affect Ca and Mg homeostasis, in particular, the low Ca and Mg reserves at birth and frequent concurrent illness with therapies that affect Ca and Mg metabolism (for example, chronic diuretic and steroid therapies with their potential for affecting renal excretion, intestinal absorption and bone metabolism of Ca and Mg).[11,12,16,17] Concurrent illnesses in many preterm infants also may delay clinical tolerance of adequate enteral feeding. Thus, preterm infants frequently are at risk for disturbances in Ca and Mg homeostasis. This chapter aims to present an overview of the disturbances in Ca and Mg homeostasis in preterm infants as manifested by changes in serum concentrations and altered skeletal mineralization.

II. CHANGES IN SERUM CONCENTRATIONS OF CALCIUM AND MAGNESIUM

Postnatal changes in serum concentrations of Ca and Mg follow a similar but often exaggerated course compared to infants born at term (see Chapter 4).

A. HYPOCALCEMIA

At birth, serum total and ionized Ca concentrations are higher in the cord blood compared to maternal values.[18,19] In the immediate newborn period, the rate and extent of decrease in serum total and ionized Ca concentrations vary inversely with birth weight and gestational age. Serum ionized Ca concentrations are positively correlated with gestational age.[18,20] In preterm infants with a very low birth weight (VLBW, <1500 g), more than 50% may have serum total Ca concentration decreases to <7 mg/dl, the traditional level regarded as hypocalcemic.[12,21,22] This marked decrease in serum Ca concentration may occur within 12 hours after birth with a proportional decrease in serum ionized Ca concentration.[22] Other factors that may predispose the infant to the development of hypocalcemia such as birth asphyxia, maternal insulin-dependent diabetes or parathyroid disorders operative in the term infant, can also result in development of hypocalcemia in the preterm infant.[12,23]

Late neonatal hypocalcemia occurring after the first few days is uncommon in preterm infants and may have the same predisposing causes as in term infants (see Chapter 4). The most common cause is from phosphate imbalance as a result of high phosphorus content of parenteral nutrition solutions.[24] In theory, the risk of neonatal hypocalcemia may be greater if there is pre-existing maternal vitamin D deficiency.

1. Clinical Manifestations

Neonatal hypocalcemia in preterm infants may be easily confused with other neonatal disorders such as hypoglycemia, sepsis, meningitis, anoxia, intracranial bleeding, and narcotic withdrawal. The less mature the infant, the more subtle and varied are the clinical manifestations. Clinical presentation may be non-specific and include apnea, cyanosis, lethargy, feeding intolerance, and abdominal distention. Frank convulsions, as seen more commonly with late neonatal hypocalcemia and in older infants, can occur at serum ionized Ca concentration of <2.5 mg/dl.[25] However, many preterm infants may remain clinically asymptomatic at the same low ionized Ca concentrations,[22] and preterm infants with frank convulsions also may have simultaneous cardiorespiratory impairment and intracranial bleeding. The problem of interpretation is also compounded by the fact that these earlier clinical reports utilized earlier generations of ionized Ca electrodes which gave lower values of ionized Ca measurements. Confirmation of hypocalcemia as the cause of clinical symptomatology includes the reversibility of clinical signs when serum Ca concentration has been increased to the normal range. Other investigations, including the measurement of parathyroid hormone, calcitonin and vitamin D metabolites, are similar to the diagnostic workup of hypocalcemia in term infants.[12,23]

2. Therapy

Prior to treatment, any neonate with suspected disturbed Ca concentrations should have blood drawn for diagnostic workup of hypocalcemia. Acute symptomatic hypocalcemia can be treated most expeditiously with intravenous infusion of 1 to 2 ml of 10% Ca gluconate or 0.3 ml to 0.5 ml of 10% Ca chloride/kg over 10 to 15 minutes, followed by continuous intravenous infusion of 30 to 75 mg elemental Ca/kg/d, usually in the form of Ca gluconate added to dextrose or dextrose electrolyte solution. Asymptomatic hypocalcemia can be treated with 30 mg to 75 mg elemental Ca/kg/d given intravenously or enterally, depending on whether the infant is tolerating adequate enteral feeding. Hypocalcemia unresponsive to usual therapy may be secondary to hypomagnesemia which should be treated accordingly (see Section B.2 below). Serial monitoring of serum Ca concentrations is essential in "titrating" the doses needed for management.[12,23]

3. Prevention

As most small infants are unlikely to tolerate adequate enteral feeding in the immediate newborn period, early use of parenteral nutrition containing adequate amounts of Ca[11] may be useful to prevent hypocalcemia. Alternately, Ca supplementation intravenously or enterally in the above doses should be started prophylactically. Serum Ca concentration should be measured every 6 to 8 hours during the first 24 hours and thereafter at less frequent intervals until parenteral nutrition or enteral feedings are started and serum total or ionized Ca concentrations are stabilized.

B. HYPOMAGNESEMIA

Serum Mg concentrations in the immediate newborn period usually remain stable.[20,22] Infants receiving low Mg intake may develop serum Mg concentrations in the traditional hypomagnesemic range (<1.5 mg/dl).[26] Hypomagnesemia beyond the first few days after birth for term infants also can have similar predisposing causes, for example, maternal insulin-dependent diabetes and maternal preeclampsia with fetal intrauterine growth retardation.[27,28] Insufficient Mg content of parenteral nutrition solutions, especially in the presence of increased losses from gastrointestinal fistula,[26] is probably the most common cause of late onset neonatal hypomagnesemia.

1. Clinical Manifestations

Clinical signs and symptoms of hypomagnesemia are usually non-specific and may be indistinguishable from hypocalcemia. Hypomagnesemia also may be asymptomatic. Confirmation of hypomagnesemia as the cause of clinical symptomatology includes the reversibility of clinical signs when serum Mg concentration has been increased to the normal range. Mg content of gastrointestinal fistula fluids can be measured, and its replacement may be necessary in the management of hypomagnesemia.

2. Therapy

Any neonate with suspected disturbed Mg concentrations should have blood drawn for diagnostic workup of hypomagnesemia prior to treatment. Acute symptomatic hypomagnesemia can be treated with intravenous infusion of 0.05 ml to 0.1 ml of 50% Mg sulfate(24 to 49 mg elemental Mg)/kg added to dextrose or dextrose saline solution and administered over 10–15 minutes. Mg also can be administered intramuscularly at the same dose. Asymptomatic hypomagnesemia can be treated with 5 mg to 10 mg elemental Mg/kg/d given intravenously or enterally, depending on whether the infant is tolerating adequate enteral feeding. Hypomagnesemia may be associated with hypocalcemia, and serum Ca concentration also should be measured and treated as necessary. In general, however, hypomagnesemia with hypocalcemia should be treated primarily with Mg, which may correct both, whereas treatment with Ca alone may compound hypomagnesemia through adverse effects on intestinal Mg absorption or urinary Mg losses. Serial monitoring of serum Mg concentrations is essential in "titrating" the doses needed for management.[12,23]

3. Prevention

Preterm infants without abnormal losses rarely develop hypomagnesemia. The use of routine parenteral nutrition and enteral feeds containing Mg[11] should be adequate for the prevention of hypomagnesemia. In the presence of large amounts of gastrointestinal fistula loss or renal loss, appropriate Mg replacement is needed.

C. HYPERCALCEMIA

Hypercalcemia (serum total Ca >11 mg/dl or ionized Ca >5.6 mg/dl) occurs much less frequently than hypocalcemia. In preterm infants, especially VLBW infants, hypercalcemia is most commonly associated with nutritional therapy. A relative deficiency of phosphate supply as in exclusively human milk-fed small preterm infants[29] or during the use of inappropriately low (or absent) phosphorus-containing parenteral nutrition solutions[24] are common causes of hypercalcemia in preterm infants. Phosphate deficiency and hypophosphatemia theoretically decrease bone formation and simultaneously lead to increased circulating 1,25(OH)$_2$D concentration with attendant increase in intestinal Ca and P absorption. In the absence of adequate phosphate supply, the increase in Ca absorbed theoretically cannot be deposited in the bone and contributes to hypercalcemia. Other causes are rare and potentially are the same as those listed for term infants (see Chapter 4).

1. Clinical Manifestations

Preterm infants with hypercalcemia may be asymptomatic. Routine screening because of known predisposing factors such as monitoring during nutritional therapy or family history of disturbed Ca metabolism may lead to its diagnosis.

Clinical signs of hypercalcemia when present are usually non-specific and include lethargy, irritability, polyuria, feeding intolerance, constipation, dehydration, and failure to thrive. Nephrocalcinosis primarily associated with the use of potent diuretics[30] theoretically might be worsened when there is co-existing hypercalcemia. Anatomic anomalies, for example, congenital heart disease associated with hypercalcemic syndromes, may be present in affected infants. Serious complications of chronic hypercalcemia such as hypertension are not well defined in preterm infants.

2. Therapy

Therapy of hypercalcemia includes management of specific underlying causes, for example, phosphate replacement in phosphate-deficient states. Non-specific treatment is usually the mainstay of therapy. For short-term treatment of acute hypercalcemic episodes, one or more courses to expand the extracellular fluid compartment with 10 ml to 20 ml of 0.9% sodium chloride/kg intravenously over one hour followed by an intravenous bolus injection of a potent loop diuretic such as 2 mg of furosemide/kg may be effective. Care should be taken to avoid fluid and electrolyte imbalance with careful monitoring of fluid balance and serum Ca, Mg, sodium and potassium concentrations, and osmolality; any imbalance should be corrected as necessary at regular intervals. Other drugs and hormone therapy are rarely needed in the treatment of hypercalcemia in preterm infants.[12,23] General treatment for chronic hypercalcemia includes dietary restriction of Ca and vitamin D and, in those infants discharged home, minimal exposure to sunlight to lower any endogenous vitamin D production.

D. HYPERMAGNESEMIA

Hypermagnesemia (serum Mg >2.5 mg/dl) may result from a combination of excessive Mg load and a relatively low capacity for renal excretion of Mg. The latter may contribute to a greater extent in preterm infants and asphyxiated infants in the development of hypermagnesemia. Hypermagnesemia most commonly occurs after maternal Mg sulfate administration for treatment of preeclampsia.[31,32] In preterm infants, serum Mg concentrations also can be easily raised to hypermagnesemic levels with an increased intake from parenteral nutrition solutions.[24,26] The use of Mg containing antacids or enemas also can cause hypermagnesemia.[12,23]

1. Clinical Manifestations

Preterm infants with hypermagnesemia may be asymptomatic. Routine screening resulting from known predisposing factors, such as monitoring during nutritional therapy or maternal history of preeclampsia and Mg sulfate therapy, may lead to its diagnosis. Hypermagnesemia-associated neuromuscular depression with floppiness, lethargy, and hyporeflexia may manifest as an exaggeration of the normal hypotonia in the very small preterm newborn. Feeding intolerance may be present and can be mistaken as an exaggerated manifestation of very small preterm newborn infants undergoing adaptation to extrauterine environment. The occurrence of electrocardiographic and other life-threatening complications of hypermagnesemia are theoretically possible but, fortunately, rare. In older preterm infants, hypermagnesemia associated with manipulation of parenteral Mg intake is usually asymptomatic. Serum Ca concentrations may be normal, decreased, or increased in hypermagnesemic infants.[12,23]

2. Therapy

Therapy for most cases of hypermagnesemia in preterm infants is to treat the underlying cause, for example, remove or decrease the source of Mg load and await spontaneous resolution with renal excretion of excessive Mg load. This is typical for the treatment of hypermagnesemia in the newborn secondary to maternal Mg therapy. However, if the clinical condition warrants more aggressive treatment, intravenous Ca given in the same dosage as for the treatment of acute hypocalcemia may be useful for acute treatment since Ca is a direct antagonist of Mg. Loop diuretic and adequate fluid intake may hasten Mg excretion. Exchange blood transfusion with citrated blood is an effective treatment (removal of Mg) for severely depressed hypermagnesemic infants. Peritoneal dialysis and hemodialysis may be considered in refractory patients. Supportive measures such as cardiorespiratory assistance and adequate hydration are needed.[12,23]

III. VITAMIN D METABOLISM

A. VITAMIN D STATUS

In humans, 25 hydroxyvitamin D (25 OHD) appears to be the major vitamin D metabolite involved in transplacental passage. It is theoretically possible that, because of shortened gestation, vitamin D stores in preterm infants may be lower than in term infants. Serum 25 OHD concentration may[33] or may not[34] increase with gestational age in neonates. Most studies show that serum 25

OHD concentration in newborn infants correlates with, but is slightly lower than, maternal serum 25 OHD concentrations.[35] It would appear that vitamin D status as indicated by serum 25 OHD concentrations in the infant is mostly determined by maternal vitamin D status which, in turn, is determined by race, sun exposure, and dietary intake rather than gestational age at birth.[35]

B. INTESTINAL VITAMIN D ABSORPTION

Early studies were interpreted that vitamin D may not be well absorbed in the intestine by preterm infants; however, serum 25 OHD concentration increased above the rachitic level of 10 ng/ml by one month of age in all preterm infants in spite of daily vitamin D intakes as low as 108 IU.[36] In a more recent study, preterm infants given daily 400 versus 800 IU of vitamin D per day had serum 25 OHD values much above the rachitic range in both groups.[34] Finally, human milk-fed preterm and term infants given daily doses of 1000 IU of vitamin D_3 have a similar increase in serum 25 OHD concentration which supports the concept of similar vitamin D absorption for preterm and term infants.[37]

C. VITAMIN D PRODUCTION IN SKIN

There are no studies of vitamin D production in skin in preterm infants. Hospitalized preterm infants should not be expected to have any significant skin production of vitamin D due to a lack of exposure to ultraviolet B in neonatal intensive care environments. Therefore, preterm infants must rely exclusively upon their vitamin D stores at birth and exogenous (nutritional) sources after birth.

D. 25 HYDROXYLATION OF VITAMIN D

Although there are no direct studies of 25 hydroxylation of vitamin D in preterm infants, the studies in which reported serum 25 OHD increase after administration of vitamin D in preterm infants[34,36,37] would support the concept that both absorption and 25 hydroxylation of vitamin D were adequate.

E. 1-HYDROXYLATION OF 25 OHD

Serum concentrations of $1,25(OH)_2D$ are usually higher in preterm than in term infants and indicate more than adequate 1-hydroxylation capabilities in preterm infants.[38,39] The highest values are seen in infants with rickets and/or fractures.[39] It is usually perceived that mineral deficiency seems to adequately stimulate $1,25(OH)_2D$ production in preterm infants.[38] With adequate Ca and phosphorus replenishment, serum $1,25(OH)_2D$ concentration promptly returns to the normal range.

F. ACTION OF $1,25(OH)_2D$

In an *in vitro* study of $1,25(OH)_2D$ action on lymphocytes, Ravid et al.[40] suggested that neonates have end organ resistance to $1,25(OH)_2D$ relative to adults. However, in their study, there was no difference in lymphocyte response between term and preterm infants.

In a study by Venkataraman et al.,[21] 19 preterm infants of less than 32 weeks gestation were given, in a non-randomized but dose-determining sequential fashion, doses of $1,25(OH)_2D$ varying from 0.1 μg/kg to 3 μg/kg. In all infants, serum Ca decreased at 24 hours and increased again by 72 hours. The authors of that study concluded that preterm infants were somewhat refractory to $1,25(OH)_2D$ since 0.68 μg/d is sufficient to elicit a calcemic response in adults with renal failure (i.e., without endogenous $1,25(OH)_2D$ production[41]). This study was not controlled for infants not receiving $1,25(OH)_2D$.

Koo et al.[22] randomized preterm infants to $1,25(OH)_2D$ at 4 μg/kg versus no medication. There was a striking difference between treated infants and controlled infants in terms of total Ca. The dose of $1,25(OH)_2D$ used in the latter study was 200 times higher than that of about 0.02 μg/kg which is sufficient to elicit a calcemic response in adults. However, no lower dose was used compared to controls. Thus, although it is possible that relative end organ resistance to $1,25(OH)_2D$ exists in very preterm infants, this requires additional investigation.

G. DO PRETERM INFANTS WITH RICKETS HAVE VITAMIN D DEFICIENCY?

A study by Koo et al.[39] showed that mean serum 25 OHD concentrations between two months and one year after birth in small preterm infants with and without rickets or fractures were similar between groups and well within the normal range. Evans and collaborators,[42] in a study of 81 small preterm infants randomized within the first few weeks after birth to receive daily supplementation of 400 IU versus 2000 IU of vitamin D, did not find any differences in terms of radiographic bone outcome by group at 2 months. Thus, vitamin D deficiency does not seem to play a major role in osteopenia or prematurity.[16]

H. VITAMIN D AND TOTAL PARENTERAL NUTRITION

Koo et al.[43] showed that parenteral doses of vitamin D as low as 25 IU/day allow maintenance of serum 25 OHD concentrations above 10 ng/ml. Thus, doses of 300-400 IU/day given currently are probably more than sufficient to maintain adequate vitamin D status, especially when vitamin D is given parenterally.[11] Also, in an *in vitro* study, Koo et al. determined that in-line filters of 0.22 μm did not reduce availability of vitamin D infused intravenously.

IV. SKELETAL MANIFESTATIONS OF DISTURBED MINERAL HOMEOSTASIS

Disturbed Ca homeostasis can affect skeletal mineralization *in utero* and postnatally and may manifest as rickets and osteopenia in the fetus and infant.[12] Isolated disturbances in Mg metabolism rarely can be associated with skeletal abnormalities, and the exact mechanisms involved have not been defined.

True fetal or congenital rickets is rare. Newborn infants of mothers with severe nutritional osteomalacia associated with Ca and vitamin D deficiency,[45,46]

Table 1 **Risk factors for the development of osteopenia and rickets in infants**

In Utero
Severe maternal nutritional osteomalacia (i.e., Ca and vitamin D deficiency)
Maternal hypoparathyroidism and hyperparathyroidism
Prolonged maternal Mg or phosphate treatment
Postnatal
Nutritional
Prolonged exclusive human milk feeding
Macrobiotic diet
Soy formula given to preterm infants
Prolonged total parenteral nutrition with low Ca and low P
Chronic loop diuretic therapy given to preterm infants
Aluminum contamination?
Inherited defects
Renal tubular disorder
Vitamin D or parathyroid hormone metabolism disorder

From Koo, W. W. K. and Tsang, R. C., *Neonatology: Pathophysiology and Management of the Newborn, 4th ed.*, J. B. Lippincott, Philadelphia, 1993, p. 585. With permission.

hyper-[47] and hypo-[48] parathyroid disorders, and prolonged treatment with Mg, for example, during preeclampsia[49,50] or with phosphate enema,[51] have been reported to have osteopenia or rickets (Table 1). The pathophysiology of disturbed fetal bone mineralization, especially in relation to maternal Mg and phosphate therapy, is poorly defined.

Postnatally, rickets and osteopenia presenting in infancy occur most frequently in small preterm infants and may occur in more than 30% of extremely low birth weight (<1 kg) infants.[52-54] The rate of occurrence depends on the nutrient intake and is associated most frequently with prolonged intake of soy formula, human milk, and low Ca and low phosphorus-containing parenteral nutrition. The common underlying cause in preterm infants appears to be mineral deficiency, particularly Ca and phosphorus, whereas in term infants, there is a relative lack of mineral intake in addition to vitamin D deficiency.[12,16,17] Isolated nutritional deficiency of copper and ascorbic acid has been reported in preterm infants with clinical and radiographic manifestations similar to rickets.[12,16] Chronic diuretic therapy and, increasingly, the use of multiple courses of corticosteroid therapy in infants with bronchopulmonary dysplasia and contamination of nutrients with toxins such as aluminum are added risk factors.[55-59] The extent, however, to which specific risk factors are responsible for the development of osteopenia and rickets is difficult to define in individual critically-ill infants receiving multiple therapies and suboptimal nutritional support. Acquired and hereditary forms of rickets that develop despite adequate availability of vitamin D usually are associated with renal tubular disorders and metabolic defects in vitamin D and parathyroid hormone metabolism. These causes of rickets are rare, and their skeletal manifestations usually do not present before late infancy (Table 1).

A. DIAGNOSIS

Pre-existing maternal illness or severe nutritional disturbance may heighten the awareness of the potential risk for skeletal abnormalities in the preterm infant. However, most cases of rickets and osteopenia are diagnosed incidentally during the investigation of complications such as fractures or conditions unrelated to the skeleton.[53,54] The presence of osteopenia and rickets is confirmed by classic radiographic features such as generalized bone demineralization and widening, cupping, and fraying of the distal metaphyses. Classic features of rickets such as severe skeletal deformities, including kyphoscoliosis and bowing of the legs, may not be present, if the diagnosis is made in early infancy before significant growth and weight bearing have occurred. This is particularly true for the preterm infant whose skeletal problem typically is diagnosed between 2 and 6 months postnatally.[16,53,54] Single measurement of biochemical values such as serum alkaline phosphatase activity or phosphorus concentrations is not predictive of underlying bone mineralization.[60] However, serial biochemical changes, including persistently low serum inorganic phosphate and serum alkaline phosphatase activity more than five times the normal adult range, may be helpful in the diagnosis.[61,62] Measurement of serum 25 OHD as an indicator of vitamin D status also may be helpful. The research use of photon absorptiometry[63,64] and, more recently, the possible use of dual energy x-ray absorptiometry[65-67] allow a more accurate quantitation of the degree of bone mineralization. Serial measurements of bone mineral content, when available, may be useful to monitor the progress of bone mineralization.

B. TREATMENT AND PREVENTION

Rickets and fractures from nutritional deficiencies respond well to adequate nutrient intake with high Ca and phosphorus content.[11,16,38] Specific therapies are required for inherited renal tubular disorders and for disorders of vitamin D and PTH metabolism.

Osteopenia, rickets, and fractures in preterm infants appear to have become less frequent since the widespread use of high Ca and high phosphorus-containing parenteral nutrition solutions[24,68,69] and formulas designed specifically for preterm infants.[11] A total daily intake of 400 IU vitamin D appears adequate to maintain normal vitamin D status in both preterm and term infants.[11,39,42] Preterm infants receiving mother's milk require supplementation with commercially available powder or liquid fortifier or liquid Ca and phosphorus preparation.[11]

The infant's mineral intake is monitored by maintaining a normal serum Ca, phosphorus, and alkaline phosphatase activity, while avoiding hypercalciuria (>6 mg Ca/kg/day). Measurement of vitamin D metabolites, such as 25 OHD and $1{,}25(OH)_2D$, and measurement of bone mineralization with standard skeletal radiographs or more precise measurements of bone mineralization with photon absorptiometry or dual energy x-ray absorptiometry, when available, may document pathologic bone mineralization.

C. OUTCOME

Short-term follow-up of infants with nutritional bone disease has shown no major residual physical deformity, and skeletal maturation, as assessed by ossification centers of the wrists for preterm infants, is similar to term infants at one year of age.[54] However, linear growth may be impaired in infants with radiographic[70] or biochemical (increased serum alkaline phosphatase activity)[71] changes of nutritional bone disease. There is also a lack of information on the long-term follow-up of these infants with regard to long-term bone mineralization and linear growth.

REFERENCES

1. Leroyer-Alizon, E., David, L., Anast, C. S., and Debois, P. M., Immunocytological evidence for parathyroid hormone in human fetal parathyroid glands, *J Clin Endocrinol Metab,* 52, 513, 1981.
2. Rubin, L. P., Posillico, J. T., Anast, C. S., and Brown, E. M., Circulating levels of biologically active and immunoreactive intact parathyroid hormone in human newborns, *Pediatr Res*, 29, 201, 1991.
3. Saggese, G., Baroncelli, G. I., Bertelloni, S., and Cipolloni, C., Intact parathyroid hormone levels during pregnancy, in healthy term neonates and in hypocalcemic preterm infants, *Acta Paediatr Scand*, 80, 36, 1991.
4. Wolfe, H. J., DeLellis, R. A., Volkel, E. F., and Tashjan, A. H., Distribution of calcitonin-containing cells in the normal neonatal thyroid gland: a correlation of morphology with peptide content, *J Clin Endocrinol Metab*, 41, 1076, 1975.
5. Samaan, N. A., Anderson, G. D., and Adam-Mayne, M. E., Immunoreactive calcitonin in the mother, neonate, child and adult, *Am J Obstet Gynecol*, 121, 622, 1975.
6. David, L., Salle, B. L., Putet, G., and Grafmeyer, D. C., Serum immunoreactive calcitonin in low birth weight infants. Description of early changes; effect of intravenous calcium infusion; relationships with early changes in serum calcium, phosphorus, magnesium, parathyroid hormone, and gastrin levels, *Pediatr Res*, 15, 803, 1981.
7. Hillman, L. A., Hoff, N., Walgate, J., and Haddad, J. G., Serum calcitonin concentrations in premature infants during the first 12 weeks of life, *Calcif Tissue Int*, 34, 470, 1982.
8. Chase, L. R. and Slatopolsky, E., Secretion and metabolic efficacy of parathyroid hormone in patients with severe hypomagnesemia, *J Clin Endocrinol Metab*, 38, 363, 1974.
9. Rude, R. K., Oldham, S. B., and Singer, F. R., Functional hypoparathyroidism and parathyroid hormone end-organ resistance in human magnesium deficiency, *Clin Endocrinol*, 5, 209, 1976.

10. Fatemi, S., Ryzen, E., and Flores, J., Effect of experimental human magnesium depletion on parathyroid hormone secretion and 1,25 dihydroxyvitamin D metabolism, *J Clin Endocrinol Metab*, 73, 1067, 1991.
11. Koo, W. W. K. and Tsang, R. C., Calcium, magnesium, phosphorus, and vitamin D, in *Nutritional Needs of the Preterm Infant: Scientific Practice and Practical Guidelines*, Tsang, R. C., Lucas, A., Uauy, R., and Zlotkin, S., Eds., Williams & Wilkins, New York, 1993, 135.
12. Koo, W. W. K. and Tsang, R. C., Calcium and magnesium homeostasis, in *Neonatology: Pathophysiology and Management of the Newborn*, 4th ed., Avery, G. B., Ed., J. B. Lippincott, Philadelphia, 1993, 585.
13. Abrams, S. A, Yergey, A. L., Schanler, R. J., Vieira, N. E. and Welch, T. R., Hypercalciuria in premature infants receiving high mineral-containing diets, *J Pediatr Gastroenterol Nutr,* 18, 20, 1994.
14. Hillman, L. S., Johnson, L. S., Lee, D. Z., Vieira, N. E., and Yergey, A. L., Measurement of true absorption, endogenous fecal excretion, urinary excretion, and retention of calcium in preterm infants by using a dual-tracer, stable-isotope method, *J Pediatr*, 123, 444, 1993.
15. Koo, W. W. K. and Tsang, R. C., Calcium and magnesium metabolism, in *CRC Handbook of Clinical Chemistry*, Werner, M., Ed., CRC Press, Inc., Boca Raton, FL, 1989, 51.
16. Koo, W. W. K. and Tsang, R. C., Bone mineralization in infants, *Prog Food Nutr Sci*, 8, 229, 1984.
17. Koo, W. W. K., Parenteral nutrition-related bone disease, *J Parenter Enteral Nutr*, 16, 386, 1992.
18. Tsang, R., Abrams, L., Joyce, T. H., III, Buncher, C. R., and Steichen, J. J., Ionized Ca in neonates in relation to gestational age, *J Pediatr*, 94, 126, 1979.
19. Wandrup, J., Kjer, J. J., Thode, J., and Siggaard-Andersen, O., The concentration of free calcium ions and total calcium in pregnancies at term, *Scand J Clin Lab Invest*, 42, 273, 1982.
20. Nelson, N. A., Finnstrom, O., and Larsson, L., Plasma ionized calcium, phosphate and magnesium in preterm and small for gestational age infants, *Acta Paediatr Scand*, 78, 351, 1989.
21. Venkataraman, P. S., Tsang, R. C., Steichen, J. J., Grey, I. G., Neylan, M., and Fleischman, A. R., Early neonatal hypocalcemia in extremely preterm infants. High incidence, early onset, and refractoriness to supraphysiologic doses of calcitriol, *Am J Dis Child*, 140, 1004, 1986.
22. Koo, W. W. K., Tsang, R. C., Poser, J. W., Laskarzewski, P., Buckley, D., Johnson, R., and Steichen, J. J., Elevated serum calcium and osteocalcin levels from calcitriol in preterm infants: a prospective randomized study, *Am J Dis Child*, 140, 1152, 1986.
23. Koo, W. W. K. and Tsang, R. C., Neonatal calcium and phosphorus disorders, in *Pediatric Endocrinology: A Clinical Guide*, 2nd ed., Lifshitz, F., Ed., Marcel Dekker, New York, 1990, 569.
24. Koo, W. W. K., Tsang, R. C., Succop, P., Krug-Wispe, S. K., Babcock, D., and Oestreich, A. E., Mineral vitamin D and high calcium and phosphorus needs of preterm infants receiving parenteral nutrition, *J Pediatr Gastroenterol Nutr*, 8, 225, 1989.

25. Sorell, M. and Rosen, J. F., Ionized calcium: Serum levels during symptomatic hypocalcemia, *J Pediatr*, 87, 67, 1975.
26. Koo, W. W. K., Fong, T., and Gupta, J. M., Parenteral nutrition in infants, *Aust Paediatr J*, 16, 169, 1980.
27. Mimouni, F., Tsang, R. C., Hertzberg, V. S., and Miodovnik, M., Polycythemia, hypomagnesemia and hypocalcemia in infants of diabetic mothers, *Am J Dis Child*, 140, 798, 1986.
28. Mimouni, F. and Tsang, R. C., Perinatal magnesium metabolism: Personal data and challenges for the 1990s, *Magnesium Res*, 4, 109, 1991.
29. Lyon, A. J., McIntosh, N., Wheeler, K., and Brooke, O. G., Hypercalcaemia in extremely low birthweight infants, *Arch Dis Child*, 59, 1141, 1984.
30. Hufnagle, K. G., Khan, S. N., and Penn, D., Renal calcifications: a complication of long term furosemide therapy in preterm infants, *Pediatrics*, 70, 360, 1982.
31. Stone, S. R. and Pritchard, J. A., Effect of maternally administered magnesium sulfate on the neonate, *Obstet Gynecol*, 35, 574, 1970.
32. Lipshitz, P. J., The clinical and biochemical effects of excess magnesium in the newborn, *Pediatrics*, 47, 501, 1971.
33. Chan, G. M., Tsang, R. C., Chen, I. W., DeLuca, H. F., and Steichen, J. J., The effect of $1,25(OH)_2$ D_3 supplementation in premature infants, *J Pediatr*, 93, 91, 1978.
34. Pittard, W. B., Geddes, K. M., Hulsey, T. C., and Hollis, B. W., How much vitamin D for neonates?, *Am J Dis Child*, 145, 1147, 1991.
35. Specker, B. L., Greer, F., and Tsang, R. C., Vitamin D, in *Nutrition During Infancy*, Tsang, R. C. and Nichols, B. L., Eds., Hanley and Belfus, Philadelphia, 1988, 264.
36. Hillman, L. S. and Haddad, J. G., Perinatal vitamin D metabolism. II Serial 25-hydroxyvitamin D concentrations in sera of term and premature infants, *J Pediatr*, 86, 928, 1975.
37. Delmas, P. D., Glorieux, F. H., Delvin, E. E., Salle, B. L., and Melki, I., Perinatal serum bone gla-protein and vitamin D metabolites in preterm and full-term neonates, *J Clin Endocrinol Metab*, 65, 588, 1987.
38. Steichen, J. J., Tsang, R. C., Greer, F. R., Ho, M., and Hug, G., Elevated serum 1,25 dihydroxyvitamin D concentrations in rickets of very low birth weight infants, *J Pediatr*, 99, 293, 1981.
39. Koo, W. W. K., Sherman, R., Succop, P., Ho, M., Buckley, D., and Tsang, R. C., Serum vitamin D metabolites in very low birth weight infants with and without rickets and fractures, *J Pediatr*, 114, 1017, 1989.
40. Ravid, A., Koren, R., Rotem, C., Amir, Y., Reisner, S., Novogrodsky, A., and Liberman, U. A., Mononuclear cells from human neonates are partially resistant to the action of 1,25-dihydroxyvitamin D, *J Clin Endocrinol Metab*, 67, 655, 1988.
41. Brickman, A. S., Coburn, J. W., and Massry, S. G., 1,25 dihydroxy-vitamin D_3 in normal man and patients with renal failure, *Ann Intern Med*, 80, 161, 1974.
42. Evans, J. R., Allen, A. C., Stinson, D. A., Hamilton, D. C., Brown, B. S., Vincer, M. J., Raad, M. A., Gundberg, C. M., and Cole, D. E. C., Effect of high-dose vitamin D supplementation on radiographically detectable bone disease of very low birth weight infants, *J Pediatr*, 115, 779, 1989.

43. Koo, W. W. K., Tsang, R. C., Steichen, J. J., Succop, P., Oestreich, A. E., Noseworthy, J., and Farrell, M. K., Vitamin D requirement in infants receiving parenteral nutrition, *J Parenter Enteral Nutr*, 11, 172, 1987.
44. Koo, W. W. K., Hollis, B. W., Horn, J., Steiner, P., Tsang, R. C., and Steichen, J. J., Stability of vitamin D_2, calcium, magnesium, and phosphorus in parenteral nutrition solution: effect of in-line filter, *J Pediatr*, 108, 478, 1986.
45. Russell, J. G. B. and Hill, L. F., True fetal rickets, *Br Radiol*, 47, 732, 1974.
46. Zhou, H., Rickets in China, in *Rickets*, Glorieux, F. H., Ed., Raven Press, New York, 1991, 253.
47. Hanukoglu, A., Chalen, S., and Kowardski, A. A., Late onset hypocalcemia, rickets and hypoparathyroidism in an infant of a mother with hyperparathyroidism, *J Pediatr*, 112, 751, 1988.
48. Gradus, D., Le Roith, D., Karplus, M., Zmora, E., Grief, M., and Bar-Ziv, J., Congenital hyperparathyroidism and rickets: Secondary to maternal hypoparathyroidism and vitamin D deficiency, *Isr J Med Sci*, 17, 705, 1981.
49. Lam, C. I., Norton, K. I., Murphy, R. J. C., Wilkins, I. A., and Rabinowitz, J. G., Congenital rickets associated with magnesium sulfate infusion for tocolysis, *J Pediatr*, 113, 1078, 1988.
50. Holcomb, W. J. Jr., Shackelford, G. D., and Petrie, R. H., Magnesium tocolysis and neonatal bone abnormalities: a controlled study, *Obstet Gynecol*, 78, 611, 1991.
51. Rimensberger, P., Schubiger, G., and Willi, U., Connatal rickets following repeated administration of phosphate enemas in pregnancy: a case report, *Eur J Pediatr*, 151, 54, 1992.
52. Callenbach, J. C., Sheehan, M. B., Abramson, S. J., and Hall, R. T., Etiologic factors in rickets of very low-birth-weight infants, *J Pediatr*, 98, 800, 1981.
53. Lyon, A. J., McIntosh, N., Wheeler, K., and Williams, J. E., Radiological rickets in extremely low birthweight infants, *Pediatr Radiol*, 17, 56, 1987.
54. Koo, W. W. K., Sherman, R., Succop, P., Krug-Wispe, S., Tsang, R. C., Steichen, J. J., Crawford, A. H., and Oestreich, A. E., Fractures and rickets in very low birth weight infants: Conservative management and outcome, *J Pediatr Orthop*, 9, 326, 1989.
55. Venkataraman, P. S., Han, B. K., Tsang, R. C., and Daugherty, C. C., Secondary hyperparathyroidism and bone disease in infants receiving long term furosemide therapy, *Am J Dis Child*, 137, 1157, 1983.
56. Hahn, T. J., Steroid and drug-induced osteopenia, in *Primer on the Metabolic Bone Diseases and Disorders of Mineral Metabolism*, 2nd ed., Favus, M. J., Ed., Raven Press, New York, 1993, 250.
57. Koo, W. W. K., Kaplan, L. A., Horn, J., Tsang, R. C., and Steichen, J. J., Aluminum in parenteral nutrition solution - sources and possible alternatives, *J Parenter Enteral Nutr*, 10, 591, 1986.
58. Koo, W. W. K., Kaplan, L. A., and Krug-Wispe, S. K., Aluminum contamination of infant formulas, *J Parenter Enteral Nutr*, 12, 170, 1988.
59. Koo, W. W. K., Krug-Wispe, S. K., Succop, P., Bendon, R., and Kaplan, L. A., Sequential serum aluminum and urine aluminum:creatinine ratio and tissue aluminum loading in infants with fractures and rickets, *Pediatrics*, 89, 877, 1992.
60. Ryan, S. W., Truscott, J., Simpson, M., and James, J., Phosphate alkaline phosphatase and bone mineralization in preterm neonates, *Acta Paediatr*, 82, 518, 1993.

61. Martinasek, D. J. and Tsang, R. C., Hypophosphatemia and bone disease in the neonatal intensive care unit, in *Phosphate in Pediatric Health and Disease*, Alon, U. and Chan, J. C. M., Eds., CRC Press, Boca Raton, 1993, 115.
62. Koo, W. W. K., Succop, P., and Hambidge, K. M., Serum alkaline phosphatase and serum zinc concentrations in preterm infants with rickets and fractures, *Am J Dis Child*, 143, 1342, 1989.
63. Mimouni, F. and Tsang, R. C., Bone mineral content: Data analysis, *J Pediatr*, 113, 78, 1988.
64. Greer, F. R., Determination of radial bone mineral content in low birth weight infants by photon absorptiometry, *J Pediatr*, 113, 213, 1988.
65. Chan, G. M., Performance of dual energy X-ray absorptiometry in evaluating bone, lean body mass, and fat in pediatric subjects, *J Bone Miner Res*, 7, 369, 1992.
66. Venkataraman, P. S. and Ahluwalia, B. W., Total bone mineral content and body composition by X-ray densitometry in newborns, *Pediatrics*, 90, 767, 1992.
67. Braillon, P. M., Salle, B. L., Brunet, J., Glorieux, F. H., Delmas, P. D., and Mennier, P. J., Dual energy X-ray absorptiometry measurement of bone mineral content in newborns: Validation of the technique, *Pediatr Res*, 32, 77, 1992.
68. MacMahon, P., Blair, M. E., Treweeke, P., and Kovar, I. Z., Association of mineral composition of neonatal intravenous feeding solutions and metabolic bone disease of prematurity, *Arch Dis Child*, 64, 489, 1989.
69. Prestridge, L. L., Schanler, R. J., Shulman, R. J., Burns, P. A., and Laine, L. L., Effect of parenteral calcium and phosphorus therapy on mineral retention and bone mineral content in very low birth weight infants, *J Pediatr*, 122, 761, 1993.
70. Koo, W. W. K., Sherman, R., Succop, P., Oestreich, A. E., Tsang, R. C., Krug-Wispe, S. K., and Steichen, J. J., Sequential bone mineral content in small preterm infants with and without fractures and rickets, *J Bone Miner Res*, 3, 193, 1988.
71. Lucas, A., Brooke, O. G., Baker, B. A., Bishop, N., and Morley, R., High alkaline phosphatase activity and growth in preterm neonates, *Arch Dis Child*, 64, 902, 1989.

Chapter 4

Neonatal Mineral Metabolism

Francis Mimouni and Winston W.K. Koo

CONTENTS

0-8493-4613-4/95/$0.00+$.50

I. DYNAMIC CHANGES IN POSTNATAL MINERAL METABOLISM

A. PHYSIOLOGIC DECREASE IN SERUM CA

The newborn infant has to overcome the challenge of switching from a "parenteral" (through the umbilical vein) to an "enteral" mineral intake. Intrauterine bone mineralization is optimized by a very peculiar hormonal situation in the fetus: Ca is transported across the placenta in an active fashion,[1] against a concentration gradient,[2] and at an impressive rate (during the last trimester of pregnancy) of up to 150 mg/kg fetal weight.[3] Serum Ca concentration is high in the fetus (around 11 mg/dl) relative to the mother (9–9.5 mg/dl).[2] High fetal serum Ca concentrations lead to decreased parathyroid hormone (PTH)[4] and 1,25 dihydroxycholecalciferol (1,25 $(OH)_2$ D) concentrations,[5,6] and to increased serum calcitonin (CT)[7,8] and 24,25$(OH)_2$D[6] concentration. Abundant substrate (Ca and phosphorus), increased bone mineralization [due to calcitonin[9] and, possibly, to 24,25$(OH)_2$D] and decreased bone resorption [due to low PTH[11] and 1,25$(OH)_2$D[12,13]] create an ideal situation in the rapidly growing fetus.

The "parenteral" infusion of Ca abruptly ceases at birth and neonatal serum ionized Ca (iCa) concentration begins to drop, reaching a nadir at 24 hours.[4,5] Initially, the rate of decrease is very rapid, as half of it is reached by 2 hours of life.[14] Decreasing serum iCa concentration, presumably due to continued mineral accretion in bone, rapidly triggers PTH secretion,[4,5] and 1,25$(OH)_2$D production;[5] the concentrations of these 2 hormones rise rapidly, and their combined effect on bone resorption (stimulatory) and on bone formation (inhibitory) leads to a stabilization of serum iCa concentration.[4,5] At the same time, enteral Ca supply is being established (through milk feeding), and, after the initial fall, serum iCa concentrations slowly rise to reach stable childhood values by 1 week of age.[14] Together with the drop of iCa concentration on day 1 of life, a very significant increase (rather than decrease) of serum CT is observed.[7,8] This increase is a paradox, as it does not follow the usual rules of feedback regulation of CT production by serum Ca.[15] It is usually interpreted as a consequence of the "stress" of birth, as CT is considered to be one of the hormones released during stress.[16] Also, CT elevation occurs during a period when enteral feeds are started, i.e., when known CT secretagogues such as enteroglucagon or gastrin are being secreted.[16]

B. POSTNATAL INCREASE IN SERUM MG CONCENTRATION

Fetal serum Mg concentrations (as estimated by neonatal cord blood values) are higher than maternal concentrations,[17] which suggest an active placental transport mechanism similar to calcium. At the placenta, the magnesium and Ca transporter system seem to be independent, as neither maternal hypermagnesemia nor maternal dietary Mg deficiency appear to affect the rate of placental Ca transport.[18,19] Studies using the *in situ* perfused rat placenta also support the existence of an active transport mechanism for Mg.[20] It would be logical that, as happens to serum Ca concentrations, serum Mg concentrations would also fall after birth, when the newly born infant is disconnected from the placenta. Furthermore, since the major route of Mg excretion is the kidney,[21] the neonatal increase in glomerular filtration rate[22] would be expected to lead to an increase in urinary Mg losses. Paradoxically, however, serum Mg concentrations *increase* over the first few days of life.[23] This phenomenon may, in part, be due to contraction of the extracellular space which occurs during the postnatal physiologic loss of weight, but is not entirely explained by it.

Mg and Ca metabolism are tightly interrelated. In acute experiments, CT lowers, PTH increases serum Mg concentrations, and Mg mimics the effects of Ca on CT and PTH secretion.[21] However, in chronic Mg deficiency, both PTH secretion and PTH action are impaired, presumably because of decreased Mg-dependent adenyl-cyclase activity.[21] In neonates, the increase in serum PTH which follows birth correlates with neonatal serum Mg concentrations,[23] and adequate Mg status appears to be important in the maintenance of neonatal serum Ca concentrations.[24]

II. EARLY NEONATAL HYPOCALCEMIA

A. CAUSES

Early neonatal hypocalcemia typically occurs during the first few days after birth, with the lowest concentrations of serum calcium being reached at 24 to 48 hours after birth. Early neonatal hypocalcemia occurs primarily in preterm infants. The frequency of hypocalcemia varies inversely with birth weight and gestational age, and the nadir of the serum calcium concentration may occur at less than 12 hours in preterm infants.[14,25,26] Hypocalcemia also occurs in infants who have suffered birth asphyxia[27] and in infants of mothers with insulin-dependent diabetes.[28] Neonatal hypocalcemia usually is more severe and prolonged when associated with maternal hyperparathyroidism (see Section III.B), fetal parathyroid hypoplasia or agenesis, or isolated hypoparathyroidism. The latter may be a part of DiGeorge Syndrome, which is characterized by the partial or complete absence of derivatives of the third and fourth pharyngeal pouches, and possibly the fifth pouch, and is often associated with defective development of the third, fourth, and sixth aortic arches.[29,30] Details of parathyroid disorders affecting infancy will be discussed in subsequent sections.

The mechanisms of development of neonatal hypocalcemia are varied, and each clinical condition may be associated with several factors that may contribute to hypocalcemia: for example, limited milk and calcium intake, in addition to transient, limited increase in the serum PTH concentration; elevated serum CT concentration; and, possibly, end organ resistance to $1,25(OH)_2D$. These factors appear to be primarily responsible for the development of hypocalcemia in the preterm infant.

Decreases in serum ionized calcium can occur without decreases in serum total calcium. Agents that complex calcium in the blood would be expected to decrease ionized calcium. The use of citrate-containing blood for exchange transfusion and disturbances in acid-base status can lead to hypocalcemia in these circumstances. These situations are encountered equally frequently beyond the first few days after birth and will be discussed in Section III.

B. CLINICAL MANIFESTATIONS

The clinical manifestation of neonatal hypocalcemia may be confused easily with other neonatal disorders such as hypoglycemia, sepsis, meningitis, intracranial bleeding, and narcotic withdrawal. The neonate with hypocalcemia also may be asymptomatic; the less mature the infant, the more subtle and varied are the clinical manifestations. Significant clinical signs are tremulousness, apnea, cyanosis, and seizures; infants also may be lethargic, feed poorly, vomit, and have abdominal distention. Frank convulsions are seen more commonly with late neonatal hypocalcemia. The degree of irritability of the infant does not appear to correlate with serum calcium values. The classic signs of peripheral hyperexcitability of motor nerves, carpopedal spasm, i.e., spasm of wrists and ankles, and laryngospasm, are uncommon in preterm infants. Therefore, suspicion of hypocalcemia should be confirmed with measurement of serum total and ionized calcium. Confirmation of hypocalcemia as the cause of clinical symptomatology is the reversibility of clinical signs when serum total calcium or ionized calcium has been increased to the normal range.[29,30]

C. TREATMENT

The most effective and most rapid means of elevating serum calcium concentrations is by intravenous administration of calcium salts. Seizures suspected to be caused by hypocalcemia should be treated with intravenous 10% calcium gluconate (1 ml/kg) administered over 10 minutes with constant monitoring of the heart rate. Gradual or abrupt decrease in heart rate during the infusion is an indication to slow or stop the infusion. Intravenous calcium therapy may be complicated by acute hypercalcemia; extravasation of calcium solution leads to skin sloughs, tissue necrosis, and calcification. If umbilical venous catheters are used, the tips should not be intracardiac if calcium is to be administered through them. Arterial infusion of calcium in high concentrations potentially is fraught with many dangers and should be avoided if possible. Anecdotal cases of massive sloughing of soft tissue in the area perfused by the peripheral artery receiving the infusion have been reported, and inadvertent administration into a mesenteric artery theoretically can lead to necrosis of intestinal tissues. Direct admixture with bicarbonate or phosphate solution will result in precipitation. Ten percent calcium chloride at approximately 0.3 ml/kg, which delivers similar amounts of elemental calcium as 10% calcium gluconate 1 ml/kg, also may be used with the same precautions as above. Prolonged use of calcium chloride in high doses may be associated with acidosis and probably should be avoided. After resolution of the seizures, intravenous calcium solution may be continued at a dose of 50 to 75 mg elemental calcium kg/day until the serum calcium concentrations have remained consistently in the normal range. Thereafter, the intravenous calcium solution can be reduced in stepwise fashion, i.e., 50% for 24 hours, 25% for another 24 hours, and discontinuation. With intravenous therapy, bolus infusion may be associated with a transient, slight increase in blood pH and serum phosphorus.[31] Continuous infusion probably is more efficacious than intermittent therapy, since renal loss of calcium may be greater with the latter method.[32]

Oral calcium therapy in the same dosage (50 to 75 mg elemental calcium kg/day in 4 to 6 divided doses) may be used for maintenance therapy. All calcium preparations are hypertonic, and there is a theoretical potential for precipitating necrotizing enterocolitis in infants at risk for this condition. Oral calcium preparations with a syrup base containing a high sucrose content may constitute a significant carbohydrate and osmolar load for very small infants and be associated with an increase in frequency of bowel movements; thus, these preparations probably should not be used for oral administration. Intravenous preparations can be used orally, and commonly include calcium glubionate (115 mg elemental calcium per 5 ml), calcium gluceptate (90 mg elemental calcium per 5 ml), and calcium carbonate (40 mg elemental calcium per ml).

The treatment of early neonatal hypocalcemia usually can be tapered after 2 to 3 days of therapy as the serum calcium concentrations return to normal. However, other causes of hypocalcemia as listed in Sections III.A–D may need prolonged therapy. A poor response to calcium therapy may be due to concurrent magnesium deficiency (see hypomagnesemia). In otherwise stable infants, the initiation of parenteral or enteral nutrition should maintain serum calcium in the normal ranges.

Neonatal hypocalcemia may resolve spontaneously. Thus, it is possible that asymptomatic neonatal hypocalcemia may not require treatment. Because of the major physiologic importance of calcium in all cellular systems, however, especially its role as a second messenger for the initiation of the intracellular enzymatic cascade, hypocalcemia potentially can alter important cellular functions and probably should be corrected. Treatment of asymptomatic hypocalcemia can be instituted with oral or intravenous calcium salts, using the regimen described above.

III. "LATE" NEONATAL HYPOCALCEMIA

"Late"-onset neonatal hypocalcemia (beyond 2 to 4 days of life) has 3 major recognized causes: phosphate loading, hypoparathyroidism, and hypomagnesemia.

A. PHOSPHATE LOADING

Phosphate loading was, in the past, a major cause of neonatal hypocalcemia, as infants were fed modified cow milk or evaporated milk with very high phosphorus content. The mechanism of phosphate-induced hypocalcemia is not completely understood, but is believed to be related to increased tissue Ca deposition, as well as to PTH resistance generated by the hyperphosphatemia.[33] With currently available "humanized" cow-milk derived formulae, neonatal hypocalcemia due to phosphate loading is uncommon. All cases recently published[34] or encountered by us were associated with feeding of one type of formula (Similac, Ross Laboratories), which is the formula richest in P content available in the U.S.A. Since this formula is the most commonly used in the U.S.A., and hypocalcemia is rare, we suggest that infants affected by this entity may have an unknown, transient defect in renal phosphate handling. In the United Kingdom, phosphate-induced hypocalcemia is often encountered in infants of Indian subcontinent Asian immigrants, which supports the theory that marginal vitamin D status could be an additional etiologic factor for this condition.[35]

Typically, patients with P-induced hypocalcemia are recognized because of tetany or frank seizures (which may be generalized or focal); they are hypocalcemic and often hypomagnesemic as well.[34] Their serum P concentration is very high. This situation mimics hypoparathyroidism, but can be differentiated from it by the presence of (appropriately) elevated serum PTH concentrations,[34] and by the spontaneous resolution of hypocalcemia and hyperphosphatemia when the infant is switched to a low P containing formula (such as SMA, Wyeth Laboratories or PM 60/40, Ross Laboratories, Columbus, OH).

B. HYPOPARATHYROIDISM

Hypoparathyroidism leads to hypocalcemia and, often, to hypomagnesemia as well.[21] In the absence of adequate PTH secretion, renal tubular reabsorption of phosphate is uninhibited,[36] and serum P concentrations rise sharply. Hypoparathyroidism may be associated with the other 2 elements of the DiGeorge triad (cardiovascular defects and T cell immune deficiency).[37] Because of the existence of "partial" DiGeorge syndromes,[38] careful cardiovascular examination and T cell immune function studies are indicated in any neonate with newly discovered hypoparathyroidism. The course of hypoparathyroidism in DiGeorge syndrome is not predictable. Although parathyroid gland remnants appear to hypertrophy and become functional in many patients, some will require lifelong treatment. Others may relapse after a period of remission lasting several years.[39]

When isolated, congenital hypoparathyroidism can be inherited as an X-linked[40] or autosomal dominant trait.[41] It may also be seen in association with chromosomal anomalies such as ring chromosomes (No. 16 or 18) or be sporadic.[42]

Congenital hypoparathyroidism may also be transient, lasting for a few days to a few months.[43] Initially, these patients are not initially distinguishable from the sporadic cases of permanent hypoparathyroidism. Their Ca salt and vitamin D metabolites requirements are similar. However, on follow-up examination, doses will have to be decreased progressively because of hypercalcemic tendency and, eventually, these infants are weaned from all therapy.[43] Whether they are truly cured, or in simple remission is unknown, as there are reports of relapse in later life.[43]

Finally, congenital hypoparathyroidism may be secondary to maternal hypercalcemia; it may be the first revealing "clinical sign" of the maternal condition, which may be hyperparathyroidism[44,45] or any other cause of hypercalcemia, such as familial benign hypocalciuric hypercalcemia.[46] It is, therefore, critical to measure serum Ca, P, and PTH concentrations in any mother of an infant with late-onset hypocalcemia; maternal results will have to be interpreted in view of lower calcium values in pregnancy (due to hemodilution). Congenital hypoparathyroidism secondary to maternal hypercalcemia is benign, as the neonatal hypocalcemia is usually resolved within a few days to a few weeks.[44,45]

C. MAGNESIUM DEFICIENCY

Magnesium deficiency leads to neonatal hypocalcemia through deficient PTH secretion[21,47] and end organ resistance to PTH.[21,48] It may be inherited as a primary selective defect in either intestinal Mg absorption (most likely X-linked)[49,50] or renal tubular Mg reabsorption;[51,52] it may also be secondary to maternal Mg deficiency, for instance, in mothers with poorly controlled diabetes mellitus and urinary Mg losses.[53]

Mg deficiency is very difficult to differentiate from hypoparathyroidism, as both conditions lead to hypocalcemia, hypomagnesemia, and low serum PTH concentrations. Only a positive calcemic and PTH response to Mg repletion will allow differentiation between the 2 conditions.[24] We suggest that hypocalcemic and hypomagnesemic patients should first be treated with Mg salts, as calcium therapy will confound the issue and may aggravate the Mg deficit. In case of seizures, the use of Ca salts might appear more pressing. In one study of late onset hypocalcemia, however, Mg salts alone controlled seizures more rapidly than Ca salts or phenobarbital.[54]

D. OTHER CAUSES OF LATE NEONATAL HYPOCALCEMIA

- *Post acidotic tetany.* Infants with acidosis secondary to dehydration or to renal tubular acidosis may develop significant hypocalcemia. Bone demineralization may be severe in acidosis[55] and leads to urinary mineral losses, including Ca and Mg. Hypocalcemia develops during correction of acidosis with alkali therapy,[56] and may coexist with severe Mg deficit.[57]
- *Respiratory or metabolic alkalinization.* Patients with persistent pulmonary hypertension are purposely maintained alkalotic by respiratory (hyperventilation) or metabolic (bicarbonate infusion) means, in order to cause pulmonary vascular bed relaxation.[58-60] Bicarbonate ion binds directly to Ca and may rapidly decrease serum iCa concentrations, even without a noticeable change in serum total Ca.[61]

- *Phototherapy*. In rats, phototherapy may lead to hypocalcemia apparently through decrease in melatonin secretion,[62] which leads to a decrease of corticosterone secretion and subsequent increase in bone Ca uptake.[62] In preterm infants, phototherapy was found by one study to be an additional risk factor for neonatal hypocalcemia.[63]

IV. HYPERCALCEMIA IN THE FIRST YEAR OF LIFE

A. GENERAL CONSEQUENCES OF HYPERCALCEMIA

Regardless of its cause, hypercalcemia, although at times asymptomatic, may have severe consequences.[64] Hypercalcemia can cause resistance to antidiuretic hormone, which may lead to polyuria with subsequent polydypsia and dehydration, somewhat similar to diabetes insipidus.[64] Hypertension may occur, secondary to a direct effect of Ca on blood vessels (vasoconstriction), or due to an increase in renin angiotensia activity; central nervous system effects of hypercalcemia may be vomiting, lethargy, hypotonia, or seizures, and may be due to a combination of direct effect of Ca on nerve cells and hypertensive encephalopathy. When long lasting, hypercalcemia may lead to metastatic tissue calcifications, most frequently nephrocalcinosis and kidney stones.[64]

An exception to the above complications is familial benign hypocalciuric hypercalcemia, a topic on which we will expand later.

B. PRIMARY HYPERPARATHYROIDISM

Congenital, primary hyperparathyroidism is very rarely observed.[65] It is usually related to diffuse clear cell parathyroid hyperplasia, as adenomas are generally not seen in the neonatal period;[65] it is most often sporadic, although 2 types of familial patterns have been reported: the first one is compatible with autosomal recessive inheritance;[65] the second one, autosomal dominant, is seen in families affected by the so-called "familial benign hypocalciuric hypercalcemia" (see Sections IV.I).

Hyperparathyroidism usually presents as a "failure to thrive" syndrome, with features of hypotonia, poor sucking, and constipation.[65] Bone deformities may be seen, secondary to severe demineralization. When the rib cage is affected, respiratory distress may be a prominent feature.[65] Laboratory findings include: hypercalcemia, hypophosphatemia, hyperphosphatasia, hypercalciuria and phosphaturia; serum PTH concentration is elevated, and aminoaciduria is frequent.[65] Skeletal X-ray survey shows diffuse demineralization, subperiosteal resorption in the phalanges, and possible spontaneous fractures. Maxillary tooth germs have lost their lamina densa, a sign very specific to bone resorption.[65]

Without treatment, progressive deterioration occurs which leads to death; in fact, several of the few reported cases were only discovered at autopsy.[65] Surgical treatment (subtotal parathyroidectomy) must be performed early in the course of this disorder.

C. SECONDARY HYPERPARATHYROIDISM

Congenital hyperparathyroidism secondary to maternal, chronic hypocalcemia is a very rare disorder, as only 16 cases have been reported to date.[66,67] Typically, these infants have radiographic features of primary hyperparathyroidism.[66] Neonatal hypercalcemia is not the rule, as only 4 out of the 16 reported cases had hypercalcemia.[66] It is a transient disorder that will have a spontaneous resolution when adequate postnatal mineral supply is provided.[66]

D. IDIOPATHIC INFANTILE HYPERCALCEMIA

Idiopathic infantile hypercalcemia (IIH) is a poorly understood condition. It has a curious overlap with the Williams Syndrome (WS) consisting of "elfin facies", mental retardation, and supravalvular aortic stenosis.[68] Many patients with WS suffered IIH in early infancy, and many patients with IIH are affected by the other features of WS.[68] However, there are definite instances of WS uncomplicated by IIH and of IIH without any of the other features of WS.[68]

The major difference between IIH and hyperparathyroidism is that in IIH, serum PTH concentrations are appropriately decreased by hypercalcemia;[68] also, since there is no excess PTH secretion in IIH, there is no increased phosphaturia, nor subsequent hypophosphatemia.[68] There is no specific diagnostic laboratory test of IIH since its cause is unknown; however, many of the features of IIH are mimicked by vitamin D intoxication, and we recommend obtaining serum 25(OH)D measurements in these patients as an index of vitamin D stores. Elevated serum $1,25(OH)_2D$ concentrations have been reported in some infants with IIH, but have not been found in others.[69] Decreased CT production has been suggested as a possible cause,[70] but is inconsistent with the finding of only very mild, mostly post-prandial hypercalcemia, in infants with complete thyroid agenesis,[71] in contrast to the severe hypercalcemia often encountered in IIH.[68]

The natural course of hypercalcemia in IIH is relatively predictable.[68] It will usually last for a few months to a few years, although pubertal relapses have been reported.[68] Since the hypercalcemia is essentially absorptive in nature, eradication of vitamin D, protection from sun, and Ca-poor diet including low Ca, vitamin D free infant formula (Calcilo XD, Ross Laboratories) are the mainstay of treatment. These treatments, if required for prolonged periods, will unfortunately lead to progressive demineralization.[68] Long-term complications (nephrocalcinosis, renal vascular disease, arterial stenoses, hypertension, etc.) may be severe in cases with significant and prolonged hypercalcemia. In severe cases, the use of exogenous CT might be a rational approach, as it should both decrease serum Ca concentration and prevent bone demineralization.[70]

E. SUBCUTANEOUS FAT NECROSIS (SFN)

SFN is a transient disorder of fat tissue, mostly encountered in newborn infants after direct trauma and/or birth asphyxia.[72-75] Histologically, SFN is accompanied by granulomatous infiltration. During the resolution phase of SFN, significant hypercalcemia may develop. It has been suggested that hypercalcemia in these infants is secondary to excessive $1,25(OH)_2D$ production by granulomatous cells, similar to what occurs in sarcoidosis.[76] In these patients, hypercalcemia is transient (a few weeks), absorptive in nature, and responds well to Ca and vitamin D poor diets; severe cases are well-controlled by glucocorticoid treatment.[73-75]

F. HYPERPROSTAGLANDIN E SYNDROME

Hyperprostaglandinuria is a key feature of Bartter's syndrome and its variants.[77] Seyberth et al recently reported one variant characterized apparently by fetal polyuria (leading to intrauterine polyhydramnios and prematurity), renal electrolyte wastage (in particular, massive hypercalciuria and kalliuria), hypercalcemia, and metabolic alkalosis.[77] These infants suffer episodes of fever, vomiting and diarrhea, and failure to thrive.[77] All these infants have diminished bone density and are euparathyroid. Most of these symptoms are corrected by the cyclooxygenase inhibitor, indomethacin.[69] Based upon studies using stable isotopes of Ca, there appears to be two components in this syndrome: an indomethacin-sensitive dietary contribution (absorptive hypercalcemia/hypercalciuria) and a relatively indomethacin resistant bone resorptive element.[78]

G. SEVERE INFANTILE HYPOPHOSPHATASIA

This autosomal recessive condition causes a deficient alkaline phosphatase activity in the bone, liver, and intestine.[79,80] As a consequence of failure of Ca to deposit in bone, hypercalcemia, hypercalciuria, and nephrocalcinosis develop. Severe congenital rickets is usually present. This syndrome may be actually due to the deficiency in a serum factor that activates alkaline phosphatase, as serum transfusions are accompanied by a long-lasting improvement in the biochemical and chemical features, as well as the appearance of alkaline phosphatase activity in osteoblasts and chondrocytes.[80] Pathognomonic of this disease is a striking increase in phosphoethanolamine production in urine.

H. HYPOPHOSPHATEMIC HYPERCALCEMIA

This condition is encountered in preterm infants fed unfortified breast milk[81] and is described in detail in the previous chapter by Koo et al.

I. FAMILIAL BENIGN HYPOCALCIURIC HYPERCALCEMIA

Described for the first time by Foley et al, this entity is characterized by mild hypercalcemia (free of its classic complications), combined with hypocalciuria; the hypercalcemia was not resolved by further parathyroidectomy.[46] Since then, many similar kindreds have been reported throughout the world. In this dominantly inherited syndrome, PTH secretion appears to be slightly increased or not adequately repressed by hypercalcemia;[46] the condition does not lead, however, to bone demineralization and increased risk of fractures.[46] In these patients, even after parathyroidectomy, renal tubular reabsorption of Ca is increased and is probably the cause of hypocalciuria.[46] It has been suggested that the mild degree of hypercalemia in this condition prevents the development of further parathyroid hyperplasia. Infants born to mothers affected by this condition may suffer transient congenital hypoparathyroidism (due to chronic parathyroid gland depression *in utero*), be normocalcemic, or be affected by moderate to severe primary hyperparathyroidism.[46] The latter situation is poorly understood.

J. IATROGENIC HYPERCALCEMIA

This is probably the most frequently observed cause of hypercalcemia in neonatal intensive care environments; it results from overtreatment with calcium salts. Since Ca binders (e.g., citrate from blood products, bicarbonate) may increase total but not iCa, it is important to directly measure the serum iCa concentration in this situation.

V. NEONATAL HYPERMAGNESEMIA

Hypermagnesemia in the neonate is now more frequently observed compared to the past. Previously, it was exclusively found in infants born to mothers with pregnancy-induced hypertension (PIH) treated with magnesium sulfate ($MgSO_4$) to prevent seizures.[82-84] In the past decade, $MgSO_4$ has been recognized as a potent tocolytic agent and is increasingly used to this effect.[85,86] With current protocols of treatment, maternal and, hence, fetal serum Mg concentrations may reach very high values (4–10 mg/dl). At such values, neonatal neuromuscular depression is frequent, which may lower Apgar scores.[83,87] Severe neonatal respiratory depression is rare. Based upon studies in adults, cardiovascular collapse is probably not seen when serum Mg values remain below 10–14 mg/dl.[88] Signs of neuromuscular depression do not correlate well with cord serum Mg values, but may correlate with the duration of Mg therapy to the mother.[87]

The impact of hypermagnesemia on neonatal Ca values tends to be protective. Indeed, these hypermagnesemic infants have higher serum Ca concentrations than their normomagnesemic controls in spite of parathyroid gland depression.[83] It has been suggested that Mg may exchange with Ca in bone, leading to its release from bone and subsequent elevation of blood calcium.[89] Also, in blood, Mg displaces Ca bound to albumin or to Ca salts, which leads to an increase in serum iCa concentrations.[90] It has recently been reported that infants born to mothers with prolonged tocolysis using $MgSO_4$ are at high risk for developing congenital rickets.[91] Using a rat model of *in situ* placental perfusion, we showed that this form of rickets is probably not due to decreased placental Ca transport, which appears to remain intact during maternal hypermagnesemia.[19] It is likely that the rachitic effect of chronic fetal hypermagnesemia is mediated through inhibition by Mg of osteocalcin-binding to hydroxyapatite;[92] osteocalcin is a protein synthesized by osteoblasts which plays a significant role in bone mineralization.[93]

VI. VITAMIN D DEFICIENCY RICKETS

A. EPIDEMIOLOGY OF VITAMIN D DEFICIENCY

Vitamin D deficiency is encountered mostly in developing countries where inadequate vitamin D skin production (insufficient sun exposure for sociocultural or climatologic reasons, possibly aggravated by dark skin complexion) is not corrected by vitamin D intake fortification of foodstuffs. At particularly high risk are middle eastern Bedouins,[94,95] Indian subcontinent Asians in Great Britain,[96,97] Eskimos,[98] and infants in China.[99,100]

In the U.S.A., vitamin D deficiency is rare and almost exclusively encountered in breast fed infants of African-American ancestry who did not receive regular prophylaxis.[101-103] The variable and generally low amount of vitamin D in human milk (equivalent to 12–60 IU/L)[104] makes infants very dependent on cutaneous production related to sun exposure and skin pigmentation.[105-107]

B. EFFECTS OF VITAMIN D DEFICIENCY

In the absence of vitamin D, intestinal absorption of Ca and P is decreased and secondary hyperparathyroidism develops in order to prevent hypocalcemia;[102] hyperparathyroidism raises serum Ca concentrations directly by increased bone resorption and indirectly by increased intestinal Ca absorption through maximal transformation of available 25 OH vitamin D to $1,25(OH)_2D$. However, if the deficiency of substrate 25(OH)D is very severe, serum concentrations of $1,25(OH)_2D$ end up falling.[108] In addition, hyperparathyroidism aggravates phosphopenia by decreasing renal tubular reabsorption of phosphate.[109] Mineral deficiency will lead to formation of unmineralized cartilage and bone matrix;[109] it also increases metabolism of 25(OH)D and further aggravates vitamin D deficiency. Furthermore, because the growth plate requires normal local concentrations of Ca for chondroclast maturation, the cartilage matrix cannot be resorbed, leading to enlarged growth plates (metaphyseal enlargement). In severe cases, secondary hyperparathyroidism becomes insufficient to maintain serum calcium concentrations, and hypocalcemic tetany or convulsions may develop.[111] A significant feature of rickets is muscle weakness and hypotonia, which results from a cellular deficit in high-energy phosphate-related metabolites in skeletal muscle.[112] Hypotonia combined with undermineralized bones may lead to significant deformities of spine, pelvis, and legs.[109]

C. PREVENTION AND TREATMENT

Prevention of vitamin D deficiency rickets is accomplished if the current RDAs for vitamin D are respected (300 IU/d before 6 months and 400 IU/d for the remainder of first year).[113] It is possible that lower intake would be sufficient to prevent rickets, but the documented absence of vitamin D toxicity at 300–400 IU/d does not support challenging the current RDAs.[113]

Treatment of existing rickets has traditionally been with high doses of vitamin D.[109] However, this mode of treatment includes a risk of hypercalcemia and/or hypercalciuria; it has recently been suggested that "physiologic" doses (400 IU/d) are as efficient and do not have these risks.[102] The latter mode of treatment requires, however, compliance by the parents.

REFERENCES

1. Stulc J, Stulcova B. Transport of calcium by the placenta of the rat. *J Physiol* 1986; 371:1–16.
2. Schonberger CW, Pitkin RM. Maternal-perinatal calcium relationships. *Obstet Gynecol* 1979; 53:74–76.
3. Forbes GB. Calcium accumulation by the human fetus. *Pediatrics* 1976; 57:976–977.
4. Loughead JL, Mimouni F, Ross, R, Tsang RC. Postnatal changes in serum osteocalcin and parathyroid hormone concentrations. *J Am Coll Nutr* 1990; 9:358–362.
5. Steichen JJ, Tsang RC, Gratton TL, Hamstra A, DeLuca HF. Vitamin D homeostasis in the perinatal period. 1,25 dihydroxyvitamin D in maternal, cord, and neonatal blood. *N Engl J Med* 1980; 302:315–319.
6. Hollis BW, Pittard WB. Evaluation of the total fetomaternal vitamin D relationships at term. Evidence for racial differences. *J Clin Endocrinol Metab* 1986; 59:652–657.
7. Bergman L. Studies on early neonatal hypocalcemia. *Acta Paediatr Scand* (suppl) 1974; 248:1–25.
8. Hillman LS, Rojanasathit S, Slatopolsky E, Haddad JG. Serial measurements of serum calcium, magnesium, parathyroid hormone, calcitonin and 25-hydroxyvitamin D in premature and term infants during the first week of life. *Pediatr Res* 1977; 11:739–744.
9. Messer HH, Armstrong WD, Singer L. Characteristics of calcium uptake by calcitonin-treated mouse calvaria *in vitro. Calc Tiss Res* 1974; 15:85–92.
10. Ornoy A, Goodwin D, Noff D, Edelstein S. 24,25 dihydroxyvitamin D is a metabolite of vitamin D essential for bone formation. *Nature* 1978; 276:517–519.
11. Feldman RS, Krieger NS, Tashjian AH. Effects of parathyroid hormone and calcitonin on osteoclast formation *in vitro. Endocrinology* 1980; 107:1137.
12. Brommage R, Neuman WF. Mechanism of mobilization of bone mineral by 1,25-dihydroxyvitamin D_3. *Am J Physiol* 1979; 237:E113-E120.
13. Bar-Shavit Z, Teitelbaum SL, Reitsma P, Hall A, Pegg LE, Trial J, Kahn AJ. Induction of monocytic differentiation and bone resorption by a 1,25 dihydroxyvitamin D_3. *Proc Natl Acad Sci USA* 1983; 80:5907–5911.
14. Loughead JL, Mimouni F, and Tsang RC. Serum ionized calcium concentration in the neonate. *Am J Dis Child* 1988; 142(5):516–518.

15. Navey-Many T, Raue F, Graver A, Silver J. Regulation of calcitonin gene expression by hypocalcemia, hypercalcemia, and vitamin D in the rat. *J Bone Mineral Res* 1992; 7:1233–1237.
16. Venkataraman PS, Tsang RC, Wen Chen I, Sperling MA. Pathogenesis of early neonatal hypocalcemia: studies of serum calcitonin, gastrin, and plasma glucagon. *J Pediatr* 1987; 110(4):599–603.
17. Paunier L, Girardin NE, Brioschi PA, Beguin F. Maternal-fetal relationship of extra and intracellular magnesium and potassium concentration. In: *Magnesium in Cellular Processes and Medicine.* Altura BM, Durlach J, Seelig MS (eds.), Karger, Basel: 1987; pp. 151–155.
18. Cruz ML, Mimouni F, Tsang RC, Hammond G. Effect of chronic maternal dietary magnesium deficiency on placental calcium transport. *J Am Coll Nutr* 1992; 11:87–92.
19. Mimouni F, Mughal Z, Hammond G. Placental calcium transport during acute maternal hypermagnesemia in the rat. *Am J Obstet Gynecol* 1993; 168:984–987.
20. Shaw AJ, Mughal MZ, Mohammed T, Maresh MJA, Sibley CP. Evidence for active maternofetal transfer of magnesium across the in situ perfused rat placenta. *Pediatr Res* 1990; 27:622–625.
21. Levine BS, Coburn JW. Magnesium, the mimic/antagonist of calcium. *N Engl J Med* 1984; 310:1253–1255.
22. Leake RD, Trygstad CW. Glomerular filtration rate during the period of adaptation to extrauterine life. *Pediatr Res* 1977; 11:959–962.
23. Loughead JL, Mimouni F, Tsang RC, Khoury JC. A role for magnesium in neonatal parathyroid gland function? *J Am Coll Nutr* 1991; 10(2):123–126.
24. Shaul PW, Mimouni F, Tsang RC, Specker BL. The role of magnesium in neonatal calcium homeostasis: effects of magnesium infusion on calciotropic hormones and calcium. *Pediatr Res* 1987; 22:319–323.
25. Koo WWK, Tsang RC, Poser JW, Laskarzewski P, Buckley D, Johnson R, Steichen JJ. Elevated serum calcium and osteocalcin levels from calcitriol in preterm infants. *Am J Dis Child* 1986; 140:1152–1158.
26. Nelson NA, Finnstrom O, Larsson L. Plasma ionized calcium, phosphate and magnesium in preterm and small for gestational age infants. *Acta Paediatr Scand* 1989; 78:351–357.
27. Tsang RC, Chen I, Hayes W, Atkinson W, Atherton H, Edwards N. Neonatal hypocalcemia in infants with birth asphyxia. *J Pediatr* 1974; 84:428–433.
28. Mimouni F, Tsang RC, Hertzberg VS, Miodovnik M. Polycythemia, hypomagnesemia and hypocalcemia in infants of diabetic mothers. *Am J Dis Child* 1986; 140:798–800.
29. Koo WWK, Tsang RC. Neonatal calcium and phosphorus disorders. In: *Pediatric Endocrinology: A Clinical Guide, 2nd edition.* Lifshitz F (ed.), Marcel Dekker, Inc, New York: 1990, pp. 560–611.
30. Koo WWK. Calcium and magnesium homeostasis in the newborn. In: *Pathophysiology and Management of the Newborn, 4th edition.* Avery GB et al (eds.), JB Lippincott Co, Philadelphia: 1993 (in press).
31. Venkataraman PS, Sanchez GJ, Parker MK, Altmiller D. Effect of intravenous calcium infusions on serum chemistries in neonates. *J Pediatr Gastroenterol Nutr* 1991; 13:134–138.

32. Brown DR, Salsburey DJ. Short term biochemical effects of parenteral calcium treatment of early onset neonatal hypocalcemia. *J Pediatr* 1982; 100:777–781.
33. Stamp TCB. The hypocalcaemic effect of intravenous phosphate administration. *Clinical Science* 1971; 40:55–65.
34. Venkataraman PS, Tsang RC, Greer FR, Noguchi A, Laskarzewski P, Steichen JJ. Late infantile tetany and secondary hyperparathyroidism in infants fed humanized cow milk formula: longitudinal follow-up. *Am J Dis Child* 1985; 139:664–668.
35. Watney PJM, Chance GW, Scott P, Thompson JM. Maternal factors in neonatal hypocalcemia: a study in three ethnic groups. *Br Med J* 1971; 2:432–436.
36. Slovik DM, Daly MA, Potts JT JR, Neer RM. Renal 1,25-dihydroxyvitamin D, phosphaturic, and cyclic-amp responses to intravenous synthetic human parathyroid hormone-(1–34) administration in normal subjects. *Clin Endocrinol* 1984; 20:369–375.
37. Harvey JC, Dungan WT, Elders MJ, Hughes ER. Third and fourth pharyngeal pouch syndrome, associated with vascular anomalies and hypocalcemic seizure. *Clin Pediatr* 1970; 9:496–499.
38. Barrett DJ, Ammann AJ, Cowan MJ, Fisher TJ, Stiehm ER. Clinical and immunologic spectrum of the DiGeorge Syndrome. *J Clin Lab Immunol* 1981; 6:1–6.
39. Gidding SS, Minciotti AM, Langman CB. Unmasking of hypoparathyroidism in familial DiGeorge syndrome by challenge with disodium edetate. *N Engl J Med* 1988; 319:1589–1591.
40. Peden VH. True idiopathic hypoparathyroidism as a sex linked recessive trait. *Am J Hum Genetics* 1960; 12:323–337.
41. Barr DGD, Prader A, Esper U, Rampini S, Marrian VJ, Forfar JO. Chronic hypoparathyroidism in two generations. *Helv Paediatr Acta* 1971; 26:507–521.
42. Tsang RC, Noguchi A, Steichen JJ. Pediatric parathyroid disorders. *Ped Clin North Amer* 1979; 26(1):223–249.
43. Bainbridge R, Mughal Z, Mimouni F, and Tsang RC. Transient congenital hypoparathyroidism: How transient is it? *J Pediatr* 1987; 3(6):866–868.
44. Kaplan EL, Burrington JD, Klementschitsch P, Taylor J, Deftos L. Primary hyperparathyroidism, pregnancy and neonatal hypocalcemia. *Surgery* 1984; 96:717–722.
45. Better OS, Levi J, Grief E, Tuma S, Gellei B, Erlik D. Prolonged neonatal parathyroid suppression: a sequel to asymptomatic maternal hyperparathyroidism. *Arch Surg* 1973; 106:722–724.
46. Marx SJ, Fraser D, Rapoport A. Familial hypocalciuric hypercalcemia: mild expression of the gene in heterozygotes and severe expression in homozygotes. *Am J Med* 1985; 78:15–22.
47. Anast CS, Mohs JM, Kaplan SL, Burns TW. Evidence for parathyroid failure in magnesium deficiency. *Science* 1972; 177:606–608.
48. MacIntyre I, Boss S, Troughton VA. Parathyroid hormone and magnesium homeostasis. *Nature* 1963; 198:1058–1064.
49. Paunier L, Radde IC, Kooh SW. Primary hypomagnesemia with secondary hypocalcemia in an infant. *Pediatrics* 1968; 41:385–402.
50. Friedman M, Hatcher G, Watson L. Primary hypomagnesemia with secondary hypocalcemia in an infant. *Lancet* 1967; 1:703–705.

51. Bar RS, Wilson HE, Mazza Ferri EL. Hypomagnesemia-hypocalcemia secondary to renal magnesium wasting. *Annals of Internal Medicine* 1975; 82:646–649.
52. Zelikovic I, Dabbagh S, Friedman AL, Goebzer ML, Chesney RW. Severe renal osteodystrophy without elevated serum immunoreactive parathyroid hormone concentrations in hypomagnesemia due to renal magnesium wasting. *Pediatrics* 1987; 79:403–409.
53. Atchley DW. On diabetic acidosis: detailed study of electrolyte balance following withdrawal and re-establishment of insulin therapy. *J Clin Invest* 1933; 12:297–334.
54. Turner TL, Cockburn F, Forfar JO. Magnesium therapy in neonatal tetany. *Lancet* 1977 (Feb. 5):283–284.
55. Green J, Kleeman CR. Role of bone in regulation of systemic acid base balance. *Kidney Int* 1991; 39:9–26.
56. Cochran M, Wilkinson R. Effect of correction of metabolic acidosis on bone mineralization rates in patients with renal osteomalacia. *Nephron* 1975; 15:98–110.
57. Frisch LS, Mimouni F. Hypomagnesemia following correction of metabolic acidosis: a case of hungry bones. *J Am Coll Nutr* 1993; in press.
58. Bifano E, Kavey RE, Pergolizzi J, Slagle T, Bergstrom W. The cardiopulmonary effects of Ca infusion in infants with persistent pulmonary hypertension of the newborn. *Pediatr Res* 1989; 25:262–265.
59. Fike CD, Hansen TN. The effect of alkalosis on hypoxia-induced pulmonary vasoconstriction in lungs of newborn rabbits. *Pediatr Res* 1989; 25:383–388.
60. Ferrara B, Johnson DE, Chang PN, Thompson TR. Efficacy and neurologic outcome of profound hypocapneic alkalosis for the treatment of persistent pulmonary hypertension in infancy. *J Pediatr* 1984; 105:457–461.
61. Hughes WS, Aurbach GD, Sharp ME, Marx SJ. The effect of the bicarbonate anion on serum ionized calcium concentration in vitro. *J Lab Clin Med* 1984; 103:93–103.
62. Hakanson DO, Bergstrom WH. Prevention of light-induced hypocalcemia by melatonin. In: Vitamin D, chemical, biochemical and clinical endocrinology of calcium metabolism. Norman AW (ed.), Walter deGruyer & Co., Berlin, New York: 1982; pp. 1163–1165.
63. Romagnoli C, Polidori G, Cataldi L, Tortorolo G, Segni G. Phototherapy induced hypocalcemia. *J Pediatr* 1979; 94(5):815–816.
64. Levine MM, Kleeman CR. Hypercalcemia: pathophysiology and treatment. *Hospital Practice* 1987; pp. 73–90.
65. Goldbloom RB, Gillies DA, Prasad M. Hereditary parathyroid hyperplasia: a surgical emergency. *Pediatrics* 1972; 49:514–523.
66. Loughead JL, Mughal Z, Mimouni F, Tsang RC, and Oestreich AE. The spectrum and natural history of congenital hyperparathyroidism secondary to maternal hypocalcemia. *Am J Perinatol,* 1990; 7(4):350–355.
67. Glass EJ, Barr DGD. Transient neonatal hyperparathyroidism secondary to maternal pseudohypoparathyroidism. *Arch Dis Child* 1981; 56:565–568.
68. Morris CA, Demsey SA, Leonard CO, Dilts C, Blackburn BL. Natural history of Williams Syndrome: physical characteristics. *J Pediatr* 1988; 113:318–326.

69. Garabedian M, Jocoz E, Guillozo H, Grimberg R, Guillot M, Gagnadoux MF, Broyer M, Lenoir G, Balsan S. Elevated plasma 1,25-dihydroxyvitamin D concentrations in infants with hypercalcemia and an elfin facies. *N Engl J Med* 1985; 312:948–952.
70. Culler FL, Jones KL, Deftos LJ. Inpaired calcitonin secretion in patients with Williams syndrome. *J Pediatr* 1985; 107:720–723.
71. Carey DE, Jones KL, Parthemore JG, Deftos LJ. Calcitonin secretion in congenital non-goitrous cretinism. *J Clin Invest* 1980; 65:832–895.
72. Sharlin DN, Koblenzer P. Necrosis of subcutaneous fat with hypercalcemia. *Clin Pediatr* 1970; 9:290–294.
73. Veldhuis JD, Kulin HE, Demers LM, Lambert PW. Infantile hypercalcemia with subcutaneous fat necrosis: endocrine studies. *J Pediatr* 1979; 95:460–462.
74. Thomsen RJ. Subcutaneous fat necrosis of the newborn and idiopathic hypercalcemia. *Arch Dermatol* 1980; 116:1155–1158.
75. Mimouni F, Merlob P, Metzker A, Reisner SH. Supraventricular tachycardia: the icebag technique may be harmful in newborn infants. *J Pediatr* 1983; 103(2):337.
76. Finne PH, Sanderrud J, Aksnes L, Bratlid D, Aarskog D. Hypercalcemia with increased and unregulated 1,25-dihydroxyvitamin D production in a neonate with subcutaneous fat necrosis. *J Pediatr* 1988; 112:792–794.
77. Seyberth HW, Rascher W, Schweer H, Kuhl PG, Mehls O, Schärer K. Congenital hypokalemia with hypercalciuria in preterm infants: a hyperprostaglandinuric tubular syndrome different from Bartter syndrome. *J Pediatr* 1985; 107:694–701.
78. Shoemaker L, Welch TR, Bergestrom W, Abrams SA, Yergey AL, Vieira N. Calcium kinetics in the hyperprostaglandin E syndrome. *Pediatr Res* 1993; 33:92–93.
79. Teree TM, Klein LR. Hypophosphatasia: clinical and metabolic studies. *J Pediatr* 1968; 72:41–50.
80. Whyte MP, Magill HL, Fallon MD, Herrod HG. Infantile hypophosphatasia: normalization of circulating bone alkaline phosphatase activity followed by skeletal remineralization. *J Pediatr* 1986; 108:82–88.
81. Greer FR, Steichen JJ, Tsang RC. Calcium and phosphate supplements in breast milk-related rickets. *Am J Dis Child* 1982; 136:581–583.
82. Cruikshank DP, Pitkin RM, Reynolds WA, Williams GA, Hargis GK. Effects of magnesium sulfate treatment on perinatal calcium metabolism. I. Maternal and fetal responses. *Am J Obstet Gynecol* 1979; 134:243–249.
83. Donovan EF, Tsang RC, Steichen JJ, Strub RJ, Chen IW, Chen M. Neonatal hypermagnesemia: effect on parathyroid hormone and calcium homeostasis. *J Pediatr* 1980; 96:305–310.
84. McGuinness GA, Weinstein MM, Cruikshank DP, Pitkin RM. Effects of magnesium sulfate treatment on perinatal calcium metabolism. II. Neonatal responses. *Obstet Gynecol* 1980; 56:595–600.
85. Cholst IN, Steinberg SF, Tropper PJ, Fox HE, Segre GV, Bilezikian JP. The influence of hypermagnesemia on serum calcium and parathyroid hormone levels in human subjects. *N Engl J Med* 1984; 310:1221–1225.
86. Elliot JP. Magnesium sulfate as a tocolytic agent. *Am J Obstet Gynecol* 1983; 147:277–284.

87. Rasch DK, Huber PA, Richardson CJ, L'Hommedien CS, Nelson TE, Reddi R. Neurobehavioral effects of neonatal hypermagnesemia. *J Pediatr* 1982; 100:272–276.
88. Randall RE, Cohen MD, Spray CC, Rossmeisl EC. Hypermagnesemia in renal failure. Etiology and toxic manifestations. *Ann Inter Med* 1964; 61:73–78.
89. MacManus J, Heaton FW. The influence of magnesium on calcium release from bone in vitro. *Biochim Biophys Acta* 1970; 215:360–367.
90. Liu CL, Mimouni F, Ho M, Tsang R. In vitro effects of magnesium on ionized calcium concentration in serum. *Am J Dis Child* 1988; 142:837–838.
91. Holcomb WL, Shackelford GD, Petrie RH. Magnesium tocolysis and neonatal bone abnormalities: a controlled study. *Obstet Gynecol* 1991; 78:611–614.
92. Wians FH, Kzech KE, Hanshka PV. Effects of magnesium and calcium on osteocalcin adsorption to hydroxyapatite. *Magnesium* 1983; 2:83–92.
93. Boskey AL, Wians FH, Hanshka PV. The effect of osteocalcin on *in vitro* lipid-induced hydroxyapatite formation and seeded hydroxyapatite growth. *Calcif Tissue Int* 1985; 37:57–62.
94. Shany S, Biale Y, Zuili I, Yankowitz N, Berry JL, Mawer B. Feto-maternal relationship between vitamin D metabolites in Israeli Bedouins and Jews. *Am J Clin Nutr* 1984; 40:1290–1294.
95. Taha SA, Dost SM, Sedrani SH. 25-hydroxyvitamin D and total calcium: extraordinarily low plasma concentrations in Saudi mothers and their neonates. *Pediatr Res* 1984; 18:739–741.
96. Preece MA, McIntosh WB, Tomlinson S, Ford JA, Dunnigan MG, O'Riordan JLH. Vitamin-D deficiency among Asian immigrants to Britain. *Lancet* 1973; i:907–910.
97. Henderson JB, Dunnigan MG, McIntosh WB, Abdul-Motaal AA, Gettinby G, Glekin BM. The importance of limited exposure to ultraviolet radiation and dietary factors in the etiologyof Asian rickets: a risk-factor model. *Quarterly J Med* 1987; 241:413–425.
98. Mazess RB, Barden HS, Christiansen C, Harper AB, Laughlin WS. Bone mineral and vitamin D in Aleutian Islanders. *Am J Clin Nutr* 1985; 42:143–146.
99. Xiao L. Survey on infant rickets. *J Chinese Pediatr (Chinese)* 1982; 20:63–70.
100. Ho ML, Yen HC, Tsang RC, Specker BL, Chen XC, Nichols BL. Randomized study of sunshine exposure and serum 25(OH)D in breast fed infants in Beijing, China. *J Pediatr* 1985; 107:928–931.
101. Edidin DV, Levitsky L, Schey W, Dumbovic N, Campos A. Resurgence of nutritional rickets associated with breast feeding and special dietary practices. *Pediatrics* 1980; 65:232–235.
102. Venkataraman PS, Tsang RC, Buckley DD, Ho M, Steichen JJ. Elevation of serum 1,25-dihydroxyvitamin D response to physiologic doses of vitamin D in vitamin-D deficient infants. *J Pediatr* 1983; 103:416–419.
103. Krieger I. Nutritional rickets returns. *Pediatr Consult Fall* 1985; pp. 113–114.
104. Specker BL, Tsang RC. Vitamin D in Infancy and Childhood: factors determining vitamin D status. *Adv Pediatr* 1986; 33:1–22.
105. Specker BL, Valanis B, Hertzberg V, Edwards N, Tsang RC. Sunshine exposure and serum 25-hydroxyvitamin D concentrations in exclusively breast-fed infants. *J Pediatr* 1985; 107:372–376.

106. Specker BL, Tsang RC. Cyclical serum 25-hydroxyvitamin D concentrations paralleling sunshine exposure in exclusively breast fed infants. *J Pediatr* 1987; 110:744–747.
107. Specker BL, Tsang RC, Ho M, Buckley M. Seasonal differences in serum vitamin D binding protein in exclusively breast fed infants: negative relationship to sunshine exposure and 25-hydroxyvitamin D. *J Pediatr Gastroenterol Nutr* 1986; 5:290–294.
108. Garabedian M, Vainsel M, Mallet E, Guillozo H, Toppet M, Grimberg R, NGuyen TM, Balsan S. Circulating vitamin D metabolite concentrations in children with nutritional rickets. *J Pediatr* 1983; 103:381–386.
109. Barness LA. Rickets of vitamin D deficiency. In: *Nelson Textbook of Pediatrics, 12th edition.* Berhman RE, Vaughan III VC, Nelson NE (eds.), WB Saunders Company, Philadelphia, 1983: pp. 179–183.
110. Clements MR, Johnson L, Fraser DR: A new mechanism for induced vitamin D deficiency in calcium deprivation. *Nature* 1987; 325:62–65.
111. Barness LA. Tetany of vitamin D deficiency. In: *Nelson Textbook of Pediatrics, 12th edition.* Berhman RE, Vaughan III VC, Nelson NE (eds.), WB Saunders Company, Philadelphia, 1983: pp. 183–184.
112. Mize CE, Corbett RJT, Vavy R, Nunnally RL, Williamson SB. Hypotonia of rickets: asequential study by P-31 magnetic resonance spectroscopy. *Pediatr Res* 1988; 24:713–716.
113. Subcommittee on the Tenth Edition of the RDA's. Vitamin D. In: *Recommended daily allowances, 10th edition.* National Academy Press, Washington, D.C. 1989; pp. 92–98.

Chapter 5

Calcium and Magnesium in Human Lactation

Heidi J. Kalkwarf and Bonny L. Specker

CONTENTS

I. CALCIUM

A. CALCIUM CONTENT OF MILK

Lactating women lose significant amounts of calcium in breast milk. The average calcium concentration of mature human milk from women in the U. S. is approximately 280 ± 26 mg/l.[1] The calcium concentration of human milk is regulated within a relatively narrow range but changes slightly (10%) over the course of lactation. Calcium concentrations increase during the first month of lactation, and then slowly decrease after approximately 4 months of lactation returning to levels similar to that observed at 1 month.[2,3] The calcium concentration of colostrum, which is produced during the first 3 days postpartum, is approximately double that of mature milk.[4]

Dietary calcium intake does not appear to influence milk calcium concentration. Moser et al. found that milk calcium concentrations were similar between Nepalese and American lactating women despite a 50% lower dietary intake of calcium in Nepalese women.[5] Specker et al. also observed similar milk calcium concentrations among strict vegetarian women compared with

0-8493-4613-4/95/$0.00+$.50

omnivorous women, even though dietary calcium intakes in vegetarian women were 53% lower.[6] In contrast, Dagnalie et al. found that the calcium concentration of breast milk was approximately 10% lower in macrobiotic vegetarians as compared to omnivorous women.[7] The reason for this difference is not clear, though other studies of this vegetarian population report a high rate of vitamin D deficiency.[8] It may be that low dietary calcium intake in the presence of vitamin D deficiency may affect milk calcium concentration.

Discrepant results have been found regarding the effects of maternal age on the calcium content of human milk. A lower milk calcium concentration has been observed among lactating adolescents compared with lactating adult women, although the differences may be explained, at least in part, by differences in the timing of milk collections.[9] Breast milk samples from adolescents were collected closer to the last feed, which in itself was associated with lower calcium concentrations. In contrast, other investigators have found no effect of maternal age or parity on the calcium concentration of milk.[3,10] Other than the stage of lactation, no other factors that influence milk calcium concentrations have been consistently identified.

At 3 months of age, an exclusively breast-fed infant is consuming an average of 750 ml of milk per day, which results in loss of approximately 210 mg of calcium per day via breast milk.[1] Despite this large calcium drain, lactating women are able to maintain their serum calcium concentrations near those of non-lactating women.[11] The source of additional calcium to compensate for that lost in breast milk may have three possible origins: increased dietary consumption, increased efficiency of utilization and conservation, or mobilization of calcium stored in bone. Evidence for these three sources is reviewed below.

B. DIETARY CALCIUM INTAKE DURING LACTATION

Part of the calcium drain imposed by lactation may be met by increased calcium intake in the diet. One characteristic of lactation is hyperphagia, and lactating women consume more food and energy per day than their non-lactating counterparts. Abrams and Murphy estimated that the increase in energy consumption during lactation was 314 kcal/d.[12] Their estimate was based on data from the 1985 and 1986 Continuing Surveys of Food Intakes by Individuals (CSFII), which is a national probability sample of the dietary intake of individuals in the 48 contiguous states. Considering that the calcium content of the diet for women aged 19–50 years in the U.S. is 397 mg/1000 kcal, lactating women could theoretically consume an additional 125 mg/d by just consuming more food.[13] Assuming that 30% of the calcium consumed is absorbed by the intestine, then increased dietary calcium consumption could account for approximately 18% of the calcium lost in breast milk.

C. CHANGES IN CALCIUM ECONOMY

Improvements in calcium economy may be achieved by either increasing the proportion of calcium that is absorbed from the intestine or decreasing the amount excreted by the kidney. Under normal physiological conditions, parathyroid hormone (PTH) and the vitamin D endocrine system are the primary hormonal mediators that alter the utilization of calcium and maintain serum calcium concentrations within a narrow range. The active metabolite of vitamin D, 1,25-dihydroxyvitamin D [1,25$(OH)_2$D], increases the efficiency of intestinal absorption and causes renal retention of calcium and, in conjunction with parathyroid hormone, mobilizes calcium from bone. PTH also increases renal re-absorption of calcium. The effects of PTH are amplified as it promotes the 1α-hydroxylation of 25(OH)D to produce 1,25$(OH)_2$D. Based on their known roles, increases in both PTH and 1,25$(OH)_2$D would be expected during lactation to maintain serum calcium concentrations.

1. Hormonal Changes

The effects of lactation on serum PTH concentrations during human lactation are not clear. Although PTH concentrations are elevated during lactation in rats, studies in lactating women have found either no change, an increase, or even a decrease.[6,11,14-17] Three factors warrant consideration when evaluating the existing data: 1) whether the appropriate comparison group was included, 2) the dietary calcium intake of the subjects, and 3) the stage of lactation. Some researchers have found that PTH concentrations increase during pregnancy, and it is not clear how long they remain high postpartum regardless of whether a women is lactating.[11] Therefore, it is important to use postpartum controls so that the normal adaptive changes that occur postpartum are separated from the effects of lactation.

Changes in calcium homeostasis during the post weaning period may play an important role in compensating for alterations in calcium balance that occur during lactation. Kent et al. found that, although PTH concentrations did not increase during lactation, they were increased at 2 months post weaning and remained elevated at 6 months post weaning.[18] Consistent with these changes was an increased renal calcium retention post weaning. Specker et al. also found elevated PTH concentrations at 12 months postpartum in women who were still lactating or had weaned their infants as compared to non-lactating postpartum controls.[11]

Increases in serum 1,25$(OH)_2$D concentrations during lactation are seen in the rat.[19] However, the expected increase in serum 1,25$(OH)_2$D concentrations during lactation in humans has not been consistently found in all studies. Greer et al. studied 18 women over 6 months of lactation and found only slight increases in serum 1,25-dihydroxy vitamin D concentrations during this period.[17] Hillman et al. compared serum 1,25-$(OH)_2$D concentrations of 28 lactating women and 20 non-lactating postpartum women at 6 weeks postpartum and found no difference between the two groups of women.[15] However, 6 weeks of lactation may not have been a sufficient period of time to elicit the expected effects. Other possible reasons that may explain the differences in findings between studies are the type of controls that were used and failure to consider dietary calcium intake as a modifier of the hormonal responses to lactation.

To further investigate responses by the vitamin D endocrine system during lactation, Wilson et al. examined changes in total versus "free" 1,25$(OH)_2$D (determined by centrifugal ultrafiltration) and the ratio of "free" 1,25$(OH)_2$D to vitamin D binding protein (DBP). 1,25$(OH)_2$D can only exert its effects in the unbound form.[20] If the concentration of DBP and, thus, the ratio of bound to unbound 1,25$(OH)_2$D changes during lactation, then the physiological and biochemical effects of 1,25$(OH)_2$D may occur even though there is no change in total 1,25$(OH)_2$D concentrations. These investigators found that total and "free" 1,25$(OH)_2$D and DPB decreased in the first 2 weeks postpartum in lactating women, and the 1,25$(OH)_2$D concentrations were below that of control women (this may have been a response to pregnancy as postpartum controls were not included in this study for comparison). Concentrations of total 1,25$(OH)_2$D and free 1,25$(OH)_2$D increased thereafter, but were never greater than in control women by 18 weeks postpartum. DBP concentrations remained stable throughout lactation at a level slightly above that of control women. Based on these data it appears that lactation does not evoke significant increases in either total or "free" 1,25$(OH)_2$D concentrations within the first 4 months of lactation.

There are only a few studies that have investigated the effect of dietary calcium on hormonal responses to lactation. It is expected that there would be a greater 1,25$(OH)_2$D response when dietary calcium intake is sub-optimal. This hypothesis is supported by a study conducted by Specker and co-workers that found significantly higher concentrations of serum 1,25$(OH)_2$D among vegetarian lactating women compared with non-lactating vegetarian women and omnivorous lactating women.[6] Mean calcium intakes among lactating vegetarian and lactating omnivorous mothers were 486 mg/day and 1038 mg/day, respectively. It was speculated that the low calcium intake among vegetarian women resulted in an adaptive hormonal response to maximize calcium absorption. These data illustrate that the hormonal response to lactation is not constant across women and is affected by factors such as diet.

Like PTH, the vitamin D endocrine system may also be evoked after weaning. Specker et al. found no differences in serum 1,25$(OH)_2$D concentrations in women who had been breast-feeding for 6 months as compared to postpartum controls, but rather increases in serum concentrations of 1,25$(OH)_2$D were observed at 12 months postpartum in women who were still lactating or who had weaned their infants as compared to postpartum controls.[11] These findings further emphasize the need to examine the post-weaning period to obtain a full understanding of the effects of lactation on calcium homeostasis.

2. Intestinal Calcium Absorption

Most recent evidence indicates that intestinal calcium absorption does not increase during human lactation, which is consistent with the lack of evidence for an increase in serum 1,25$(OH)_2$D concentrations. Kent et al. used stable isotopes to measure fractional calcium absorption in 49 pregnant women at 36 weeks gestation and in 31 of these women after 3 months of lactation and in 26 non-pregnant, non-lactating control women.[21] Fractional calcium absorption was 24% higher in women who were pregnant as compared to controls. After 3 months of lactation, fractional calcium absorption decreased and was only 8% higher than controls (not statistically different). A similar pattern of increased fractional calcium absorption during pregnancy and decrease towards non-pregnant levels during lactation was found in a stable isotope study by Cross et al.[22] In 2 of the 3 women who were measured before and during pregnancy and during lactation, fractional calcium absorption increased during pregnancy and returned to baseline levels during lactation. All women had a constant dietary calcium intake the day prior to the absorption tests. No information was given regarding usual calcium intakes.

Specker et al. measured fractional calcium absorption in 8 women who were consuming different amounts of calcium for at least 6 weeks.[23] Each woman was studied once when lactating and once when non-lactating. No effect of lactation on calcium absorption was observed. During lactation, fractional calcium absorption was 19% for women consuming a low calcium diet (370 mg/d; n = 4) and 23% for women consuming a similar diet that was supplemented with 1500 mg calcium/d (n = 4). In these same women, studied when they were not lactating, fractional calcium absorption was 18% and 16%, respectively.[23]

In contrast to the data on fractional calcium absorption, data from a 4-day calcium balance study showed an increased net calcium absorption (i.e., intake – calcium excretion) during lactation as compared to controls.[24] The increase in net calcium absorption was similar to that seen in the last trimester of pregnancy. The discrepancy in results from balance and stable isotope studies is probably due to the small number of balance studies performed and the large variability and opportunity for error in measurements from balance studies. In addition, balance studies and stable isotope studies measure two conceptually different entities that differ in their treatment of endogenous calcium excretion. Balance studies measure apparent or net calcium absorption, and unabsorbed calcium can not be distinguished from calcium that was endogenously secreted into the gastrointestinal tract. Absorption in a balance study would appear to be increased if endogenous secretion decreased during lactation. Stable isotope studies using two isotopes, one administered orally and the other intravenously, measure true calcium absorption and do not take into account endogenous calcium secretion. Although net calcium absorption is the best measure of the *net* response of the intestine to a physiologic stressor such as lactation, net calcium absorption is more difficult to measure, more subject to error, and does not distinguish between changes in true absorption versus changes in endogenous secretion.

3. Urinary Calcium Excretion

Current research indicates that renal re-absorption of calcium may be the major site of physiologic calcium conservation during human lactation.[18,23,25] Specker et al. observed a 44% reduction in urinary calcium excretion during lactation in women consuming a low calcium diet and a 22% reduction in urinary calcium excretion in women consuming a high calcium diet.[23] Overall, urinary calcium excretion was 45 mg/d less during lactation. This could account for approximately 20% of calcium lost in breast milk. Kent and co-workers observed a 21% reduction in fasting urinary calcium excretion in 40 lactating women compared with 40 non-postpartum control women.[18] This renal calcium conservation was sustained throughout 6 months post weaning. These authors speculated that the increased renal re-absorption of calcium played an important role in the subsequent increase in bone mineral content post weaning. The mechanism for the renal conservation of calcium during lactation and post weaning is not clear. Changes in renal calcium excretion occur without concomitant changes in PTH and $1,25(OH)_2D$, the most commonly recognized regulators of renal calcium excretion.

D. MOBILIZATION OF CALCIUM FROM BONE

1. Bone Mineral Content and Density

If there were no physiologic adaptations to conserve calcium during lactation and dietary calcium intake was not increased, then a woman (with a total body calcium content of 950 g) who exclusively breast-feeds her infant for 6 months could theoretically lose 4% of her bone mass. Thus, lactation may have a substantial impact on bone, especially considering that age-related bone loss after menopause is approximately 1% per year.[26]

Several approaches have been used to evaluate the effects of lactation on bone. The three general types of studies reviewed here include 1) retrospective or cross sectional studies that have compared bone mineral content of women with different histories of lactation, 2) longitudinal studies that have measured changes in bone mineral during lactation, and 3) retrospective studies that compared the risk of osteoporotic fracture in women who breast-fed as compared to women who had not.

a. Cross-Sectional Studies

Some investigators have found that bone mass is lower in women who breast-fed children for long periods compared with women who did not breast-feed or breast-fed fewer numbers of children for shorter periods.[27,28] However, other investigators have not found this difference, and some investigators found that bone mass was actually higher in women who had breast fed.[29-36] Two factors may explain these discrepant findings: the timing of bone mass measurements and dietary calcium intake. Measurements taken during the reproductive years or shortly after lactation had ceased may be different than when bone mass measurements were taken peri- or postmenopause. It is possible that the effects of lactation on bone mass are also influenced by dietary calcium intake. If dietary calcium was sufficient to meet the demands of lactation, then bone mass may have been preserved. Conversely, if dietary calcium intake was low, then more pronounced effects of lactation on bone mineral mass could have become apparent.

Few cross-sectional studies that have examined the effect of lactation on bone mass have provided adequate information to assess calcium intake during the time women were breast-feeding. One study in which calcium intake was reported to exceed the RDA for lactating women (1200 mg/d) has found lower bone mineral content in the ultra distal forearm in women who breast-fed for more than 6 months compared with women who breast-fed for less than 6 months.[27] No change in cortical bone mineral content of the forearm was found. However, dietary calcium intake estimates were obtained after lactation had ceased. In a follow-up study by the same investigators, there were no differences in bone mineral density at the lumbar, ultra distal radius, and three hip sites among women who were nulliparous or who breast-fed 1 to 2 children, or who breast-fed three to four children.[30]

Walker et al. were able to examine whether the effects of pregnancy and lactation on bone mass differed according to dietary calcium intake by comparing 2 groups of women with different dietary habits.[37] They compared radiological measures of bone mass in Bantu and Caucasian women with large families and long lactations to women with smaller families. The average daily calcium intake among Bantu mothers was 341 mg/d and among Caucasian mothers was 609 mg/d. There were no differences in any of the indices of bone density between mothers with large and small families or between Bantu and Caucasian mothers despite low calcium intakes in the Bantus. However, it is possible that the radiographic methods for assessing bone loss may not have been sensitive enough to detect the expected differences. Also, the response to a low calcium intake may differ between Bantu and Caucasian women.

b. Longitudinal Studies

Prospective clinical studies provide more detailed information about the change in bone density during lactation. Several studies have now found that there is a significant decrease in bone mass during lactation, especially at sites that have a large proportion of trabecular bone such as the lumbar spine, femoral neck, and ultra distal radius.[38-42] Studies that have failed to find a decrease in bone mass during lactation are limited by either small sample sizes, a short follow-up period or measurements only at cortical bone sites. The magnitude of bone loss during lactation at different sites ranges from 3.5 to 6.5% at the lumbar spine, 4.8% at the femoral neck, and 2 to 7% at the distal radius. In those studies that reported dietary calcium intake, the mean intake usually met or exceeded recommended levels for lactating women of 1200 mg/d.[38,39,42] Sowers et al. found no relationship between the amount of bone lost after 6 months of lactation and dietary calcium intake or physical activity.[39] It is possible that the low circulating estrogen concentrations that accompany lactation are a predisposing factor for bone resorption.

Kent and co-workers were the first to report that the decrease in bone mineral density that occurs during lactation is accompanied by a comparable increase in bone mineral density after weaning.[18] In the last 2 years several other investigators have also found this "rebound" in bone mineral post weaning.[39,40] The subsequent increase in bone mineral density appears to compensate, at least in part, for the loss in bone density that occurs during lactation. If compensation is complete, then the calcium drain invoked by lactation should not result in a net loss in bone mass. Theoretically, this may not occur if pregnancies are closely spaced. Increases in bone mineral density that occur during the post weaning period may explain why some cross-section and retrospective studies have found no effect of lactation on bone mass or even an increase.

Despite adaptations that may occur in calcium metabolism during lactation and post weaning, lactating adolescents may have an increased requirement for calcium. Calcium requirements are high for the adolescent mother due to her own growth. The additional calcium demands that pregnancy and milk production put on the mother may need to be met by increased calcium intake. A prospective intervention study in adolescents found that the group receiving their usual intake of 900 mg calcium/d had a 10% decrease in bone mineral density at the distal forearm by 16 weeks postpartum, whereas there was no significant change in bone mineral density among lactating adolescents who received at least 1600 mg calcium.[43]

Findings by Kent et al. support the idea that additional dietary calcium can help preserve bone mineral during lactation.[25] Twenty-seven lactating, 21 pregnant and 27 non-lactating women underwent a calcium load test to evaluate the homeostatic responses in calcium metabolism after consuming 1 gram of oral calcium. At baseline, serum alkaline phosphatase was greater in the lactating group than in the control group, and osteocalcin concentrations were greater in the lactating group as compared to the both the control and pregnant groups. There were no changes in serum concentrations of these markers of bone formation in any group after the calcium load test, i.e., additional calcium did not appear to stimulate bone formation. However, this was not the case for indicators of bone resorption. At baseline, urinary hydroxyproline excretion was greater in the lactating group as compared to the control group. After the calcium load there was a significant decrease in urinary hydroxyproline excretion indicating that additional calcium depressed bone resorption. The fall in urinary hydroxyproline was greatest in lactating women. Based on these data the authors concluded that lactating women may benefit more than pregnant or non-pregnant, non-lactating women from supplemental calcium.

Calcium kinetic studies provide another measure of bone resorption and accretion. A recent study by Specker et al. found that both bone formation and resorption were increased during lactation, but the increase in bone resorption was greater.[23] The effects on bone resorption were more pronounced in women who were consuming a low calcium diet. Biochemical markers of bone turnover also indicate an increase in both bone formation and resorption during lactation. Kent et al. and Sowers et al. found increased concentrations of serum alkaline phosphatase and osteocalcin, which are biochemical markers of bone formation, in lactating women.[18,39] The decrease in bone mineral density that was observed in both studies must have been a result of a greater degree of bone resorption than formation. This would be consistent with the increase in urinary hydroxyproline excretion, a biochemical marker of bone resorption, that was observed by Kent et al.[18]

2. Risk of Osteoporotic Fracture

From a clinical standpoint, the most meaningful outcome is the effect of lactation on risk of future fracture. Epidemiologic studies have not found lactation to increase risk of osteoporotic fracture. If anything, lactation may have a protective effect, though the reduction in risk tends to be small and not statistically significant. In a prospective cohort study, Kelsey et al. found that the risk of osteoporotic fracture at the distal forearm associated with breast-feeding one child was 0.92 (95% confidence interval [CI] 0.83–1.02) and at the proximal humerus the risk was 0.88 (95% CI 0.75–1.04).[44] The length of breast-feeding that conferred this reduction in risk was not specified. In a case-control study, Alderman et al. examined the effects of cumulative months of lactation on risk of hip and forearm fracture and found an irregular, non-significant decrease in fracture risk with increasing duration of lactation.[45] However, the number of women who had previously breast-fed was small, and the study may not have had sufficient power to identify an effect. Aloia et al. found that non-osteoporotic women were twice as likely to have breast-fed their children than osteoporotic women with vertebral crush fractures.[46] Although this difference was statistically significant ($p < 0.05$), the effects of parity and other potentially confounding factors, such as age at menopause and smoking, were not equal between the two groups. In contrast to the studies cited above, Van Hemert et al. found that the risk of osteoporotic fracture was 3-fold greater in women who breast-fed for more than 12 months.[47] However, it is not clear whether the effects of age were adequately controlled in the analysis or the appropriate comparison group was used. Overall, the weight of evidence indicates that lactation per se does not increase a woman's risk of osteoporotic fracture later in life.

II. MAGNESIUM

A. MAGNESIUM CONTENT OF MILK

The concentration of magnesium in human milk averages 33 mg/l and remains relatively stable throughout lactation.[2,48] Dietary magnesium intake does not appear to affect milk magnesium concentrations. Moser et al. found similar magnesium concentrations in milk from Napalese and American women despite the fact that the Napalese women consumed 42% more magnesium.[5]

A fully breast-feeding woman producing 750 ml of milk per day loses approximately 25 mg of magnesium per day in breast milk. Over a 6-month period this represents 22% of her body stores, assuming a 60 kg woman has a total body pool size of 21 g.[49] Despite this loss of magnesium via milk, serum concentrations of magnesium are not affected. Moser et al. found that plasma and erythrocyte concentrations of lactating women were similar to those of postpartum control women and remained stable during 6 months of lactation.[50] No difference in serum magnesium concentrations between lactating and postpartum controls were found despite lactating women's consuming more magnesium. There was no relationship between dietary magnesium and plasma or erythrocyte magnesium concentrations. Furthermore, consumption of supplemental magnesium from prenatal vitamins did not increase serum magnesium concentrations, which was also true for non-lactating women. Maintenance of plasma magnesium under different dietary intakes and under conditions of magnesium loss in breast milk indicates an effective system for magnesium homeostasis.

B. CONSERVATION MECHANISMS

Increased food consumption is one way in which lactating women compensate for the magnesium lost in breast milk. Based on a magnesium density of 143 mg/1000 kcal in the diet of U.S. women between the ages of 19 and 50 years, lactating women could theoretically consume an additional 45 mg of magnesium per day. Assuming that 50% of dietary magnesium is absorbed by the intestine, then increased dietary intake could account for 90% of the magnesium lost in breast milk.[13] Thus, the magnesium body pools of lactating women who increase their food intake should, theoretically, not be severely depleted.

Like calcium, magnesium conservation during lactation is primarily achieved by renal conservation. Landkammer et al. studied 6 lactating, 6 postpartum controls (both 2–3 months postpartum) and 7 never-pregnant women participating in a magnesium balance study.[51] All women were consuming similar diets that contained 215 mg/d of magnesium, which corresponds to the typical intake of American women. Urinary magnesium excretion was 50.3 ± 2.7 mg/d in lactating women as compared to 63.2 ± 2.6 in postpartum controls and 83.2 ± 3.0 mg/d in never-pregnant women. Thus, it appears that renal conservation can compensate for approximately 52% of the magnesium lost in breast milk. Differences in urinary magnesium losses between postpartum controls and never-pregnant women may reflect a compensation to make up for losses in magnesium stores during pregnancy. No differences in apparent magnesium absorption and magnesium balance among groups were observed. Differences in magnesium balance would be expected if postpartum women were increasing body stores that were depleted during pregnancy. Failure to find differences in magnesium balance may have been due to the small sample sizes of groups.

Based on the above data on magnesium economy during lactation, it appears that healthy women can adequately adapt to the magnesium drain during lactation by increasing magnesium consumption and increasing renal magnesium retention.

REFERENCES

1. Subcommittee on Nutrition During Lactation, Committee on Nutritional Status During Pregnancy and Lactation, Food and Nutrition Board, Institute of Medicine, National Academy of Sciences, *Nutrition During Lactation*, National Academy Press, Washington D. C., 1991.
2. Karra, M. V., Kirksey, A., Galal, O., Bassily, N. S., Harrison, G. G., Jerome, N. W., Zinc, calcium, and magnesium concentrations in milk from American and Egyptian women throughout the first 6 months of lactation, *Am J Clin Nutr*, 47, 642, 1988.
3. Prentice, A., Barclay D. V., Breast-milk calcium and phosphorus concentrations of mothers in rural Zaire, *European J Clin Nutr*, 45, 611, 1991.
4. Suzuki, K. T., Tamagawa, H., Hirano, S., Kobayashi, E., Takahashi, K., Shimojo, N., Changes in element concentration and distributioin in breast-milk fractions of a healthy lactating mother, *Biol Trace Elem Res*, 28, 109, 1991.
5. Moser, P. B., Reynolds, R. D., Acharya, S., Howard, M. P., Andon, M. B., Calcium and magnesium dietary intakes and plasma and milk concentrations of Nepalese lactating women, *Am J Clin Nutr,* 47, 735, 1988.
6. Specker, B. L., Tsang, R. C., Ho, M., Miller, D., Effect of vegetarian diet on serum 1,25-dihydroxyvitamin D concentrations during lactation, *Obstet Gynecol*, 70, 870, 1987.
7. Dagnelie, P. C., van Staveren, W. A., Roos, A. H., Tunstra, L. G. M. Th., Burema, J., Nutrients and contaminants in human milk from mothers on macrobiotic and omnivorous diets, *European J Clin Nutr*, 46, 355, 1992.
8. Dagnelie, P. C., Vergote, F. J., Van Staveren, W. A., van den Berg, H., Dingjan, P. G., Hautvast, J. G. A. J., High prevalence of rickets in infants on macrobiotics diets, *Am J Clin Nutr*, 51, 202, 1990.
9. Lipsman, S., Dewey, K. G., Lonnerdal, B., Breast-feeding among teenage mothers: milk composition, infant growth, and maternal dietary intake, *J Pediatr Gastroenterol Nutr,* 4, 426, 1985.
10. Feeley, R. M., Eitenmiller, R. R., Jones, B., Barnhart, H., Calcium, phosphorus, and magnesium contents of human milk during early lactation, *J Pediatric Gastroenterol Nutr*, 2, 262, 1983.
11. Specker, B. L., Tsang, R. C., Ho, M., Changes in calcium homeostasis over the first year postpartum: effect of lactation and weaning, *Obst Gynecol*, 78, 56, 1991.
12. Abrams, B., Murphy, S., Changes in energy intakes during pregnancy and lactation in a representative sample of US women, *FASEB J*, 5, A916, 1991.
13. USDA (U. S. Department of Agriculture), *Nationwide Food Consumption Survey, Continuing Survey of Food Intakes by Individuals, Women 19-50 Years and their Children 1-5 Years, 1 Day, 1986,* Report No. 86-1, Nutrition Monitoring Division, Human Nutrition Information Service, U.S. Department of Agriculture, Hyattsville, MD, 1987.
14. Cushard, W. G. Jr., Creditor, M. A., Canterbury, J. M., Reiss, E., Physiologic hyperparathyroidism in pregnancy, *J Clin Endocrinol Metab,* 34, 767, 1972.

15. Hillman, L., Sateesha, S., Haussler, M., Wiest, W., Slatopolsky, E., Haddad, J., Control of mineral homeostasis during lactation: interrelationships of 25-hydroxyvitamin D, 24,25-dihydroxyvitamin D, 1,25-dihydroxyvitamin D, parathyroid hormone, calcitonin, prolactin, and estradiol, *Am J Obstet Gynecol,* 139, 471, 1981.
16. Retallack, R. W., Jeffries, M., Kent, G. N., Hitchcock, N. E., Gutteridge, D. H., Smith, M., Physiological hyperparathyroidism in human lactation, *Calcif Tissue Int*, 22 (suppl), 142, 1977.
17. Greer, F. R., Tsang, R. C., Searcy, J. E., Levin, R. S., Steichen, J. J., Mineral homeostasis during lactation—relationship to serum 1,25-dihydroxyvitamin D, 25-hydroxyvitamin D parathyroid hormone, and calcitonin, *Am J Clin Nutr,* 36, 431, 1982.
18. Kent, G. N., Price R. I., Gutteridge D., Smith, M., Allen, J. R., Bhagat, C. I., Barness, M. P., Hickling, C. J., Retallack, R. , Wilson, S. G., Devlin, R. D., Davies, C., St. John, A., Human lactation: forearm trabecular bone loss, increased bone turnover, and renal conservation of calcium and inorganic phosphate with recovery of bone mass following weaning, *J Bone Min Res,* 5, 361, 1990.
19. Halloran, D. P., Barthell, E. N., Deluca, H. F., Vitamin D metabolism during pregnancy and lactation in the rat, *Proc Natl Acad Sci*, 76, 5549, 1979.
20. Wilson, S. G., Retallack, R. W., Kent, J. C., Worth, G. K., Gutteridge, D. H., Serum free 1,25-dihydroxyvitamin D and the free 1,25-dihydroxyvitamin D index during a longitudinal study of human pregnancy and lactation, *Clin Endocrinol*, 32, 613, 1990.
21. Kent, G. N., Price, R. I., Gutteridge, D. H., Rosman, K. J., Smith, M., Allen, J. R., Hickling C. J., Blakeman, S. L., The efficiency of intestinal calcium absorption is increased in late pregnancy but not in established lactation, *Calcif Tissue Int*, 48, 293, 1991.
22. Cross N., Vieira, N. E., Johnson, L., Hillman, L., True fractional calcium absorption using 42Ca and 44Ca during pregnancy and lactation, *FASEB J*, 5, A1945, 1992.
23. Specker, B. L., Vieira, N. E., O'Brian, K. O., Ho, M., Heubi, J. E., Abrams, S. A., Yergey, A. L., Calcium kinetics in lactating women with low and high calcium intakes, *Am J Clin Nutr*, 1994, in press.
24. Toverud, S. U., Boass, A., Hormonal control of calcium metabolism in lactation, *Vitamins Hormones,* 37, 303, 1979.
25. Kent, G. N., Price, R. I., Gutteridge, D. H., Allen, J. R., Blakeman, S. L., Bhagat, C. I., St. John, A., Barness, M. P., Smith M., Evans, D. V., Acute effects of an oral calcium load in pregnancy and lactation: findings on renal calcium conservation and biochemical indices of bone turnover, *Miner Electrolyte Metab,* 17, 1, 1991.
26. Johnston, C. C., Hui, S. C., Wiske, P., et al., Bone mass at maturity and subsequent rates of loss as determinants of osteoporosis, in *Osteoporosis: Recent Advances in Pathogenesis and Treatment,* DeLuca, H. F., Frost, H. M., Jee, W. S. S. et al., Eds., University Park Press, Baltimore, 1981.
27. Wardlaw, G. M., Pike, A. M., The effect of lactation on peak adult shaft and ultra-distal forearm bone mass in women, *Am J Clin Nutr,* 44, 283, 1986.

28. Lissner, L. , Bengtsson, C., Hansson, T., Bone mineral content in relation to lactation history in pre- and postmenopausal women, *Calcif Tissue Int*, 48, 319, 1991.
29. Kritz-Silverstein, D., Barrett-Connor, E., Hollenbach, K. A., Pregnancy and lactation as determinants of bone mineral density in postmenopausal women, *Am J Epidemiol*, 136, 1052, 1992.
30. Koetting, C. A., Wardlaw, G. M., Wrist, spine, and hip bone density in women with variable histories of lactation, *Am J Clin Nutr*, 48, 1479, 1988.
31. Sowers, M. F., Wallace, R. B., Lemke, J. H., Correlates of forearm bone mass among women during maximal bone mineralization, *Preventive Med*, 14, 585, 1985.
32. Aloia, J. F., Caswani, A. N., Yeh, J. K., Ross, P., Ellis, K., Cohn, S. H., Determinants of bone mass in postmenopausal women, *Arch Intern Med,* 143, 1700, 1983.
33. Melton III, L. J., Bryant, S. C., Wahner, H. W., O'Fallon, W. M., Malkasian, G. D., Judd, H. L., Riggs, B. L., Influence of breastfeeding and other reproductive factors on bone mass later in life, *Osteoporosis Int*, 3, 76, 1993.
34. Hansen, M. A., Overgaard, K., Riis, B. J., Christiansen, C., Potential risk factors for development of postmenopausal osteoporosis—examined over a 12-year period, *Osteoporosis Int*, 1, 95, 1991.
35. Hreshchyshyn, M. M., Hopkins, A., Zylstra, S., Anbar, M., Associations of parity, breast-feeding, and birth control pills with lumbar spine and femoral neck bone densities, *Am J Obstet Gynecol*, 159, 138, 1988.
36. Feldblum, P. J., Zhang, J., Rich, L. E., Fortney, J. A., Talmage, R. V., Lactation history and bone mineral density among perimenopausal women, *Epidemiology*, 3, 527, 1992.
37. Walker, A. R. P., Richardson, B., Walker, F., The influence of numerous pregnancies and lactations on bone dimensions in South African Bantu and Caucasian mothers, *Clin Sci,* 42, 189, 1972.
38. Hayslip, C. C., Klein, T. A., Wray, H. L., Duncan, W. E., The effects of lactation on bone mineral content in healthy postpartum women, *Obstet Gynecol,* 73, 588, 1989.
39. Sowers, M. F., Corton, G., Shapiro, B., Jannausch, M. L., Crutchfield, M., Smith, M. L., Randolph, R. F., Hollis, B., Changes in bone density with lactation, *JAMA*, 269, 130, 1993.
40. Kalkwarf, H. J., Specker, B. L., Bianchi, D., Ranz, J., Uetrecht, G., Changes in bone mineral during lactation and post-weaning, *Am Soc Bone Mineral Res*, 8(Suppl 1), 154, 1993.
41. Krebs, N. F., Reidinger, C., Robertson A. D., Brenner M., King, J. C., Hambridge, K. M., Longitudinal measurements of bone metabolism and bone density during lactation, *FASEB J*, 6, A1083, 1992.
42. Ritchie, L. D., King, J. C., Fung, E. B.,Van Loan, M. D., Cann, C. E., Pregnancy and lactation cause spinal bone mineral loss in humans, *FASEB J,* 7, A67, 1993.
43. Chan, G. M., McMurry, M., Westover, K., Englebert-Fenton, K., Thomas, M. R., Effects of increased dietary calcium intake upon the calcium and bone mineral status of lactating adolescent and adult women, *Am J Clin Nutr,* 46, 319, 1987.

44. Kelsey, J. L., Browner, W. S., Seeley, D. G., Nevitt, M. C., Cummings, S. R., Risk factors for fractures of the distal forearm and proximal humerus, *Am J Epidemiol,* 135, 477, 1992.
45. Alderman, B. W., Weiss, N. S., Daling, J. R., Ure, C. L., Ballard, B. H., Reproductive history and postmenopausal risk of hip and forearm fracture, *Am J Epidemiol*, 124, 262, 1986.
46. Aloia, J. F., Cohn, S. H., Vaswani, A., Yeh, J. K., Yuen, K., Ellis, K., Risk factors for postmenopausal osteoporosis, *Am J Med*, 78, 95, 1985.
47. Van Hemert, A. A., Vandenbroucke, J. P., Birkenhager, J. C., Valkenburg, H. A., Prediction of osteoporotic fracture in the general population by a fracture risk score, *Am J Epidemiol,* 132, 123, 1990.
48. Dewey, K. G., Lonerdal, B., Milk and nutrient intake of breast-fed infants from 1 to 6 months: Relation to growth and fatness, *J Ped Gastroenterol Nutr*, 2, 497, 1983.
49. Shils, M. E., Magnesium, *Modern Nutrition in Health and Disease*, 7th edition, Shils, M. E., Young, V. R., Eds., Lea & Febiger, Philadelphia, 1988, 160.
50. Moser, P. B., Issa, C. F., Reynolds, R. D., Dietary magnesium intake and the concentration of magnesium in plasma and erythrocytes of postpartum women, *J Am Coll Nutr,* 4, 387, 1983.
51. Landkammer, J. L., Mangels, A. R., Moser Veillon, P. B., Magnesium metabolism: conservation mechanisms in postpartum women consuming a controlled magnesium intake, A*m J Clin Nutr,* 56, 767, 1992.

Chapter 6

Calcium and Magnesium Metabolism—Beyond Infancy

David B. Allen and Frank R. Greer

CONTENTS

0-8493-4613-4/95/$0.00+$.50

I. CALCIUM

A. NORMAL CALCIUM METABOLISM DURING CHILDHOOD

Calcium is the most abundant mineral in the body, 99% of which is located in the skeleton. A major role for calcium during childhood is to support three components of skeletal growth and maturation:

1. Growth of the skeleton in length, volume, and mass.
2. Skeletal maturation or modeling (i.e., consolidation of bone into its definitive form by enchondral ossification).
3. Remodeling and repair of bone tissue.

While the third component persists throughout life, the first and second occur only in children and adolescents. Skeletal growth velocity varies with age, approximating 25 cm in the first year, 10–15 cm the second year, and 5–6 cm per year until the prepubertal period. Growth rates accelerate to 10–12 cm per year during puberty, then decline gradually and cease entirely about four years after the end of puberty, coincident with closure of the epiphyseal plates of long bones. (Table 1) By late adolescence the bones have achieved final adult configuration. Skeletal growth is not homogeneous throughout the body. In girls, the highest rate of growth before puberty is in the lower part of the body, whereas the pubertal growth spurt occurs almost entirely in the upper part of the body.[1]

Yearly net estimates for calcium retention vary with age with minor sex and ethnic differences (Table 2). While the body of the full-term infant contains 25–30 grams of calcium, the normal young adult contains 900–1200 grams of calcium. This net retention of about 1 kilogram of calcium is reflected by age-related increase in bone mineral content (BMC) during childhood.[2] Though BMC can be determined by a number of methods, single photon absorptiometry of the distal radius (which measures the mineral in a centimeter length of bone) has been most widely used in children. Between ages 1–6 years and 6–14 years, BMC increases at a rate of 14%[3] and 10.5%[4,5] per year, respectively. During late adolescence, following the period of peak growth velocity,[6,7] the most rapid increase in bone mineral content of the skeleton occurs (Table 3). Subsequently, BMC may continue to increase well into the third decade of life.

Table 1 **Skeletal growth in the fetus, child and adolescent**

Fetus	45–50 cm
1st year	20 cm
2nd year	10 cm
3–12 years	5 cm/year
13–18 years	10 cm/year
Total	185 cm

Data from Royer (see Reference 1).

Table 2 **Net calcium retention with age**

Age in Years	Grams/Year	Milligrams/Day	Total Grams
0 (fetus)	25	Up to 125	25–30 at birth
0 to 2	70	190	165 at 2 yrs
2 to 8	20 to 25	55 to 70	285 to 315 at 8 yrs
8 to 10	50	135	385 to 415 at 10 yrs
10 to 18	80 to 100	220 to 270	860 to 1020 at 18 yrs

Data from Reference 2.

Table 3 **Radial bone mineral content in infants, children, adolescents and adults**

Birth	.090 g/cm
6 years	.350–.460 g/cm
14 years	.800–.900 g/cm
20 years	.950–1.00 g/cm
60 years	.770–1.226 g/cm

Data from References 3–7.

Requirements for calcium intake beyond infancy through adolescence remain controversial. Dietary calcium varies widely in children without obvious clinical signs of deficiency or excess. The tenth edition of the Recommended Dietary Allowances (RDA) cites a calcium intake of at least 800 mg/day for children between ages 1–10 years; however, this allowance is thought to be somewhat arbitrary since specific data on requirements of this age group are lacking.[8] Since skeletal growth in these children requires at most 190 mg per day of calcium (Table 2), an intake of 800 mg per day would clearly appear to be adequate. For adolescents, an RDA of 1200 mg per day of calcium is recommended, the increase being necessary for the calcium requirements of the pubertal growth spurt (net retention of 275–500 mg per day).[8]

These general estimates of calcium requirements, however, are influenced by numerous factors, which can be divided into several broad categories: 1) genetic and environmental factors, including general health and nutritional status, 2) hormonal factors, and 3) factors affecting calcium availability and conservation, including dietary intake, intestinal absorption, and urinary retention/excretion.

1. Genetic and Environmental Factors

Genetic constitution, which determines sex, ethnicity and ultimate skeletal size and growth, is the single most important factor affecting pediatric calcium needs and metabolism. For instance, males in general have greater bone mass than females and the BMC of the radius is about 8% larger in boys than in girls from 6–13 years of age.[9] Genetic factors affecting skeletal growth and calcium metabolism do so by directly affecting specific metabolic or enzymatic processes. In general, genetic factors are mediated by the various hormonal factors discussed below. The effects of various environmental factors potentially influencing skeletal growth and calcium metabolism are not completely defined. These include emotional and psychological factors, overall health and nutritional status including acute and chronic illnesses, and socio-economic factors. As with genetic factors, these environmental factors can have major impact on other factors discussed below.

2. Hormonal Factors

Hormonal control of calcium homeostasis involves parathyroid hormone (PTH), vitamin D, and calcitonin. Parathyroid hormone-related peptide (PTHrP), growth hormone, somatomedins, and sex steroids are also involved.

a. Parathyroid Hormone

The parathyroid glands, the principal regulators of extracellular ionized calcium (Ca^{2+}), are formed embryologically from the cells of the third and fourth pharyngeal pouches. Secretion rate of PTH, an 84-amino acid peptide, by the parathyroid chief cells varies inversely with the serum Ca^{2+} level, and is inhibited by chronic hypomagnesemia and biofeedback from 1,25-dihydroxyvitamin D ($1,25(OH)_2D$) serum concentrations. Rapid degradation of PTH yields a physiologically inactive C-terminal fragment (slowly cleared by the kidney) and an active NH_2 terminal fragment. PTH binds to cell-surface receptors in bone and kidney and exerts its effects through formation of cyclic adenosine monophosphate (cAMP). In bone, resulting increases in osteoclastic activity lead to dissolution of bone mineral and release of Ca^{2+} and PO_4 into the extra cellular fluid (ECF).

At the renal level, PTH inhibits proximal tubular PO_4 reabsorption and promotes phosphaturia. This loss of PO_4 inhibits bone mineralization and tends to shift the flow of calcium from bone to the ECF. Distal tubular reabsorption of filtered Ca is also increased by PTH. A third renal effect of PTH is the stimulation of 1α-hydroxylation of 25(OH)-vitamin D (25OHD). Metabolically active $1,25(OH)_2D$ stimulates intestinal absorption of Ca and PO_4. The

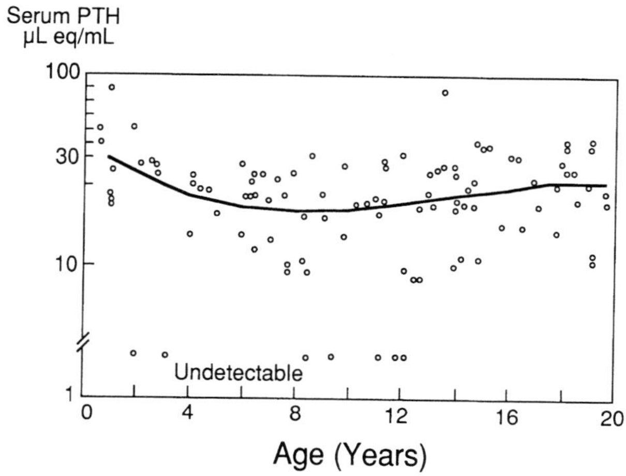

Figure 1 Serum parathyroid (PTH) concentrations over time in children and adolescents. (From Arnaud, S. B. et al., *Pediatr Res,* 7, 485, 1973. With permission.)

overall effect of PTH is to raise serum Ca levels and lower serum PO_4 levels. This characteristic reciprocal relationship is helpful in distinguishing PTH disorders from those involving vitamin D alone.

Generally, serum concentrations of PTH in children are similar to or higher than adult levels (Figure 1).[10,11] The exact values vary with age and type of PTH assay used. As in adults, the normal inverse relationship between ionized calcium and PTH exists. Also, age-dependent changes in serum PTH concentration have been noted, with the highest values reported in the youngest children and a nadir occurring at 7–9 years of age. A moderate rise occurs in early adolescence, prior to the pubertal growth spurt.[12] PTH returns to the adult normal serum value by about 20 years of age (Figure 1).[12] There is no difference between sexes until adolescence, when values in girls tend to be higher during the growth spurt compared to that in boys.[12]

b. Vitamin D

With ultraviolet irradiation, provitamin D in the skin is converted to previtamin D, which slowly isomerizes to vitamin D. A carrier protein in the blood then transports vitamin D from the skin to the liver. In the liver it is hydroxylated to 25-OHD, which is in turn hydroxylated a second time in the kidney to form $1,25(OH)_2D$. This final hydroxylation is stimulated by both decreased serum concentrations of calcium or phosphorus or increased serum PTH. $1,25(OH)_2D$ increases intestinal calcium absorption, and acting with PTH, mobilizes calcium and phosphorus from bone. It may increase urinary retention of calcium. By maintaining serum calcium and phosphorus concentrations, $1,25(OH)_2D$ maintains the milieu for promoting bone mineralization.

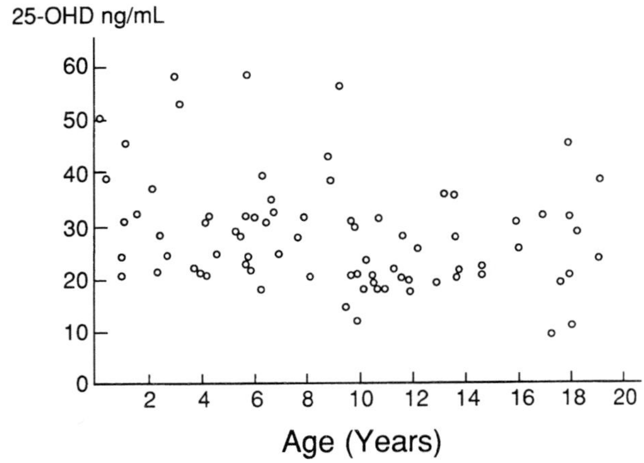

Figure 2 Serum 25-hydroxyvitamin D (25OHD) concentrations over time in children and adolescents. (From Lund, B. et al., Age dependent variations in serum 1,25-dihydroxyvitamin D in childhood, *Acta Endocrinol,* 94, 426, 1980. With permission from Scandinavian University Press, Oslo, Norway.)

The serum concentrations of 25OHD, thought to reflect overall vitamin D status, are essentially the same in children and adolescents as in adults (Figure 2).[13,14] There are no differences with increasing age after the newborn period and no sexual differences have been reported.[13,15] As with adults, seasonal differences in 25OHD have been reported, with higher serum concentrations in summer than in winter.[13,16]

Concentrations of 1,25$(OH)_2$D in children and adolescents are reported to be equal to or greater than adult values (Figure 3).[13,14,17,18] Serum concentrations have been reported to peak in adolescence in both males and females.[19,20] In children, higher concentrations of 1,25$(OH)_2$D positively correlate with skeletal growth velocity.[18] No study to date has reported gender differences with 1,25$(OH)_2$D[15] and, unlike 25OHD, there is no seasonal variation.[13,16]

c. Calcitonin

Calcitonin, produced by the C cells of the thyroid glands, has an anti-PTH effect, lowering serum calcium and phosphorus concentrations. At the level of bone, calcitonin decreases resorption of calcium and phosphorus while it has calciuric and phosphaturic effects on the kidney. However, the relative importance of calcitonin in the regulation of calcium metabolism is not clear. Thyroidectomized patients have little in the way of disturbances of calcium metabolism and relative excess or deficiency of circulating calcitonin has little effect on serum calcium concentrations. Though the exact role of calcitonin in overall calcium metabolism remains unknown, it may play some role in maintaining skeletal integrity.[21] Serum calcitonin concentrations are very high in infants and decrease with increasing age, but remain higher than adult values throughout childhood (Figure 4).[22,23]

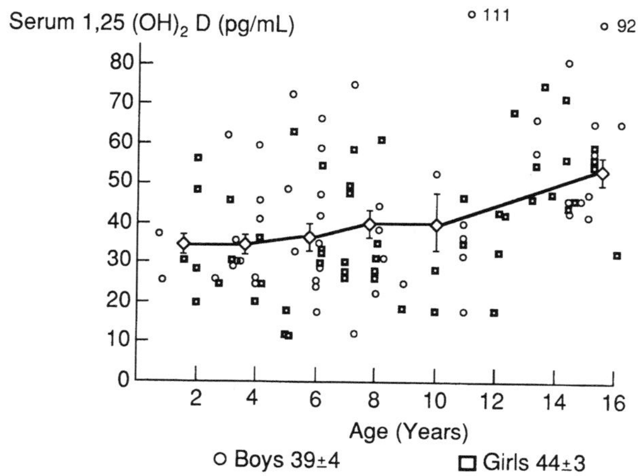

Figure 3 Serum 1, 25-dihydroxyvitaimn D (1,25[OH]$_2$D) concentrations over time in children and adolescents. (From Chesney, R. W. et al., *Am J Dis Child,* 134, 135, 1980. Copyright 1980, American Medical Association. With permission.)

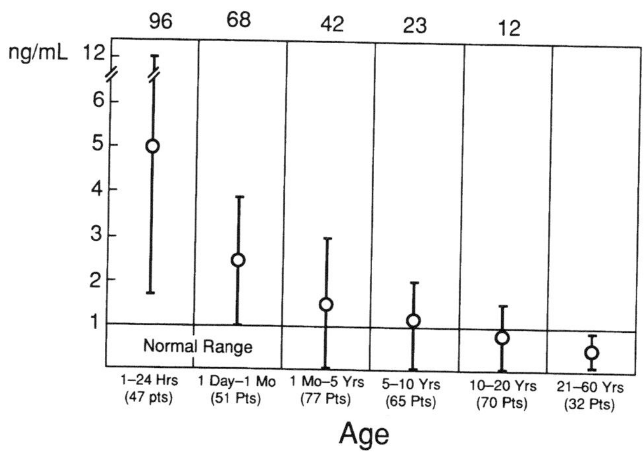

Figure 4 Serum calcitonin concentrations over time in children and adolescents. Percent of subjects with calcitonin level above normal range. (From Samaan, N. et al., *Am J Obstet Gynecol,* 121, 622, 1975. With permission.)

d. Parathyroid Hormone-Related Peptide (PTHrP)

PTHrP is a polypeptide distinct from PTH which shares a very similar 13-amino acid sequence at the N terminal end. It is responsible for elevating serum calcium levels in some patients with certain cancers (particularly breast and squamous epithelium) even without bony metastases.[24] It is also found in high concentration in human and animal breast milk.[25] While no specific function in health has been assigned to this substance, this high concentration in milk may indicate a role, perhaps unrelated to calcium metabolism, in the physiology of the infant.

e. Growth Hormone and Somatomedins

Growth hormone (GH), a secretory product of the anterior pituitary gland, is an important modulator of human growth. Both GH and GH-induced insulin-like growth factors (particularly IGF-1, previously referred to as somatomedin-C) play important roles in skeletal growth and calcium metabolism.[26,27] For example, GH-deficient children have osteopenia which improves after exogenous GH therapy (an observation that may be due, in part, to delayed skeletal maturation).[28] In these patients, bone mineral deficits observed in adults with GH-deficiency of childhood onset, appear to result from deficient build-up of bone mass during childhood and adolescence.[29] GH increases bone mass by promoting endosteal bone formation and calcium incorporation into bone.[30,31]

GH clearly influences renal handling of calcium and phosphorus, primarily through its effects on phosphorus reabsorption and calcitriol production. A well recognized effect of GH in man is stimulation of renal tubular reabsorption of phosphorus, reduction in urinary phosphorus excretion,[32] and an increase in serum phosphorus concentration.[33] GH also appears to facilitate (perhaps through IGF-1 effect) the normal compensatory increase in renal 1-α-hydroxylase activity during phosphorus deprivation,[34] thereby promoting intestinal calcium and phosphorus absorption. In spite of an inhibitory effect of GH on urinary calcium excretion, paradoxical hypercalciuria is usually observed during GH therapy and in states of GH excess. Hypercalcemia has been rarely reported in children treated with GH,[35] though this was not confirmed in subsequent studies.[36,37] Thus, important effects of GH on mineral metabolism result from its effects on the kidney.

Though the effect of endogenous GH on PTH secretion remains controversial, studies in children treated with GH have demonstrated no significant changes in serum PTH concentrations.[36,37] In children receiving GH replacement therapy, conflicting data exist regarding the effect of growth hormone on plasma $1,25(OH)_2D$ concentrations.[36-39] Overall, it would appear that there is a minimal or short-lived effect. However, significantly increased intestinal calcium absorption may occur in patients with acromegaly[40,41] and in children treated with GH.[38,41]

f. Sex Steroids

Adolescence is characterized by marked changes in serum sex steroid concentrations as well as important increases in skeletal stature and BMC of both trabecular and cortical bone. The importance of sex steroids to normal skeletal development during adolescence is illustrated by the abnormally low bone mineral content in estrogen deficient patients with Turner's syndrome[42] as well as decreased bone mass in hypogonadal males with testosterone deficiency.[43,44] In tall boys treated with testosterone, an increase in bone density has been

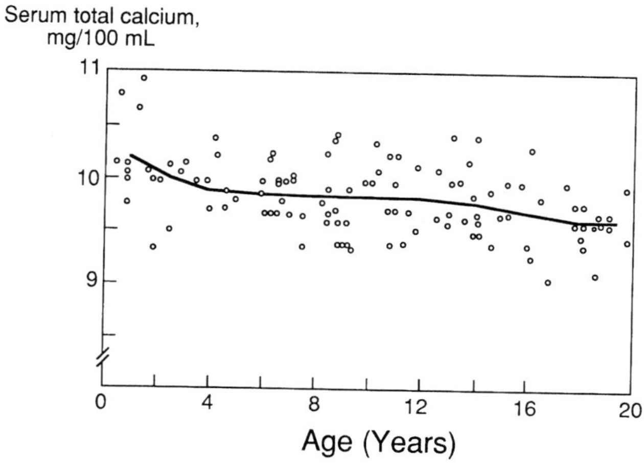

Figure 5 Total serum calcium concentrations in children and adolescents. (From Arnaud, S. B. et al., *Pediatr Res.* 7, 485, 1973. With permission.)

described.[45] Excess endogenous androgen production in young women also has been associated with increased bone density.[46]

Changes in mineral metabolism in adolescents are evidenced by early increases in serum alkaline phosphatase, serum phosphorous, and serum 1,25$(OH)_2$D.[19] Maximal serum concentration of testosterone is achieved concurrently with the maximal rise in alkaline phosphatase, and this precedes the maximal rise in BMC by five to ten months.[7] Thus, changes in mineral metabolism occur in adolescence before the greatest changes in growth velocity and bone density, providing the necessary materials for bone growth and maturation. These changes involve the pubertal surges of testosterone in males. The exact relationship between sex steroids and PTH and/or 1,25$(OH)_2$D are unknown; however, there is not a clear relationship between changes in testosterone serum consentration and 1,25$(OH)_2$D serum concentrations.[20]

B. FACTORS AFFECTING CALCIUM AVAILABILITY AND CONSERVATION

After the neonatal period, the normal values for total serum calcium are slightly higher than those of adults. Children 6 months to 2 years of age have the highest total serum calcium concentration of 10.2 mg/dl (2.54 mmol/l) which falls to 9.8 mg/dl (2.41 mmol/l) by 6–8 years and reaches the adult value of 9.6 mg/dl (2.36 mmol/l) by ages 16–20 years (Figure 5).[10] Ionized calcium values fall from 4.59 mg/dl (1.14 mmol/l) at ages 6–12 years to an adult value of 4.22 mg/dl (1.05 mmol/l) by 16–20 years.[10] Serum calcium concentrations in children are maintained by the concerted action of a number of hormones as discussed above, as well as by intestinal absorption of dietary calcium and retention of calcium by the kidney. Ultimately, dietary calcium availability

Table 4 **Absorption of dietary calcium by rats of various ages % of ca intake**

Age (weeks)	Absorbed	Re-excreted
4	98	0.7
12	57	2
24	46	6
48–72	41	7
100	24	6

Data from Reference 48.

and, thus, the nutritional status of growing children play an important role in pediatric calcium metabolism.

1. Intestinal Calcium Absorption

Most of the research on the role of the gastrointestinal tract in the control of mineral metabolism has been conducted in animals. In the rat, calcium absorption is a passive process until the time of weaning (3 weeks of age), after which time there is the development of active calcium absorption. Subsequent to weaning, the passive absorption of calcium declines but is still present until at least sexual maturity (6 weeks). The active process of calcium absorption peaks at about 35 days of age in the rat when maximum growth is occurring, after which it too decreases.[47] Calcium is 98% absorbed at 4 weeks of life, decreasing to 24% by 100 weeks (Table 4).[48]

Active duodenal calcium transport is a 1,25$(OH)_2$D dependent process. Significant concentrations of intestinal receptor for 1,25$(OH)_2$D do not appear until 18–28 days of life in the rat.[49,50] 1,25$(OH)_2$D apparently stimulates the synthesis of intestinal calcium-binding protein, the concentration of which increases dramatically at the time of weaning along with the number of 1,25$(OH)_2$D intestinal receptors.[51]

Similarly, though there is little information about the physiology and ontogeny of calcium absorption in humans, these are some general points of agreement in the literature. The efficiency of absorption is increased during periods of increased needs. Premature infants may absorb 75–80% of ingested calcium, while as little as 15–20% is absorbed by adults (Figure 6).[52-54] It is also apparent that a higher percentage of ingested calcium is absorbed at low intakes compared to high intakes. On the other hand, total amount of calcium absorbed even in sufficiency states seems to increase proportionally to intake up to a point in all age groups studied.[55,56] In a recent review of 34 published calcium balance studies in all ages, calcium balance values tend to flatten and become constant at higher calcium intakes, while at lower intakes, balance was invariably correlated with calcium intake. The "threshold values", at which balance no longer rose in response to increasing intake, exceeded the RDA for calcium for all age groups studied. This "threshold level" of calcium intake is highest

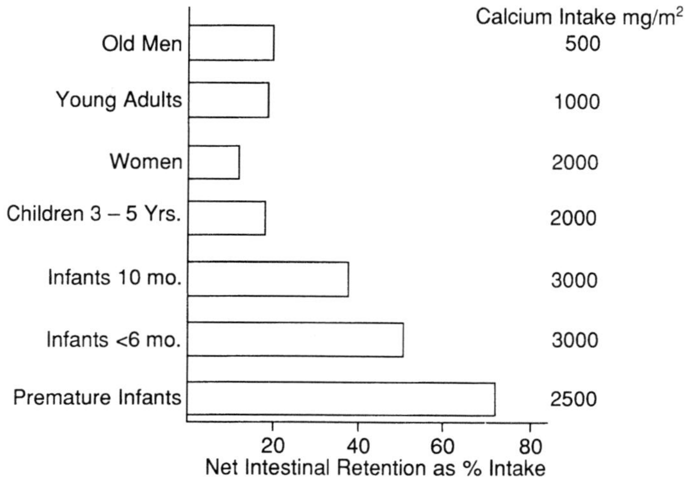

Figure 6 Percent calcium retention and calcium intake per square meter in infants, children, and adults. (From Harrison, H. E., *Fed Proc,* 18, 1090, 1959. With permission.)

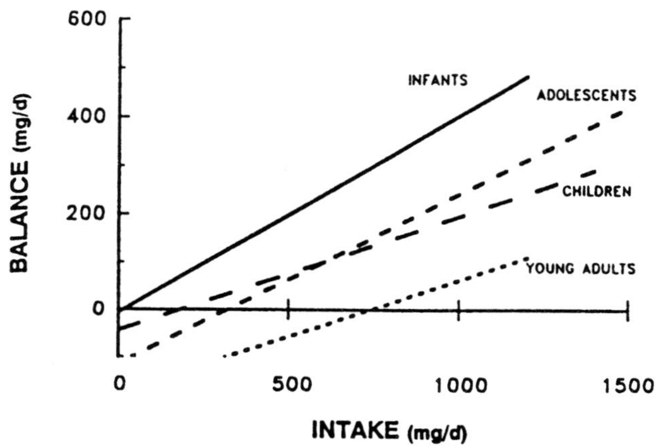

Figure 7 Graphical presentation of the regression lines for the ascending portion of the relationship of balance with intake, by age group. Highest intake for each group represents the "threshold" or plateau level (From Matkovic, V. and Heaney, R. P., *Am J Clin Nutr,* 55, 992, 1992. Copyright Robert P. Heaney, 1991. With permission.)

in infants, higher in adolescents than in younger children, and lowest in adults (Figure 7).[57]

2. Renal Regulation of Calcium Metabolism

Calcium homeostasis is largely regulated at the level of the intestine. Renal conservation of calcium is less important; however, if calcium appearing in the glomerular filtrate was not reabsorbed, relatively large urinary losses of calcium would occur. Ninety-eight percent of filtered calcium is reabsorbed by the renal tubules.[58] However, only plasma-ionized calcium and calcium complexed with various anions (60% of the total plasma calcium) is freely filterable. The mitochondria of the proximal tubule of the kidney are the location of the enzyme necessary for 1-α-hydroxylation of 25OHD to make 1,25$(OH)_2$D. The activity of the 1-α-hydroxlyase is controlled by hypocalcemic stimuli promoting PTH secretion. A direct effect of hypophosphatemia in stimulating renal 1,25$(OH)_2$D synthesis also occurs. PTH increases renal reabsorption of calcium and increases excretion of phosphate through a cyclic AMP-mediated process in the proximal renal tubule.[59] Calcitonin generally increases renal calcium excretion by the renal tubules, though the exact mechanism of action is unknown. In the vitamin D depleted state, 1,25$(OH)_2$D may increase tubular calcium reabsorption but studies are contradictory.

An early study of calcium excretion in normal subjects found that the mean calcium excretion averaged 2 mg per kg per day at all ages in children and varied little with dietary intake.[60] A later study using 24-hour urine collections to measure calcium excretion similarly reported a normal mean calcium excretion of 2.38 mg per kg per day with an upper normal limit of 4 mg per kg per day.[61]

3. Dietary Intake

More than 55% of the calcium intake of the U.S. population comes from dairy products (i.e., milk, cheese, yogurt).[62] Other dietary sources of calcium include leafy green vegetables, lime-processed tortillas, and calcium-precipitated tofu. Alternative rich sources of calcium include bones, especially the soft bones of fish and tips of poultry leg bones. Certain dietary factors may increase calcium absorption including certain amino acids (lysine, arginine) and lactose. Other foods may decrease the absorptive efficiency of calcium including cocoa, soy beans, kale, spinach, fiber containing items, and bean or wheat meal which contain hexaphosphoinosital.

The effect of inorganic phosphorus on calcium absorption remains controversial. However, excess consumption of phosphorus (by decreasing serum calcium) may contribute to the increased PTH serum concentration seen with age, and the asymptomatic, but progressive, osteopenia.[63] Large intakes of sodium phosphate containing carbonated beverages in adolescence may be of concern. However, as cola soft drinks contain only about 70 mg of phosphorus per 12 oz. can, two soft drinks would only contribute 12% of the average 1200 mg daily phosphorus intake in adolescents.[64] Perhaps a greater dietary concern regarding soft drinks is the fact that indirect evidence suggests that their intake decreases milk consumption, and thus decreases dietary calcium. A recent study in the U.S. found that only 15% of children older than 11 years of age met the RDA for calcium intake (1200 mg per day).[65]

Calcium supplements in children remain an area of controversy. In a recent report, 45 pairs of identical twins were studied in a three-year, double-blind, placebo-controlled trial of the effect of calcium supplementation (1000 mg of calcium citrate malate per day) on bone mineral density (radius, hip and spine).[66] Mean ages of the twins were 10 ± 2 years with a range of 6–14 years. In each pair, one twin served as a control for the other. The mean daily calcium intake for the twins given the placebo was 980 mg and that of the twins given the supplement was 1612 mg. Only in the prepubertal twins (n = 22 pairs) did the calcium supplementation enhance the rate of increase of bone density (4–5 % in the radius). There were no differences in the twins who went through puberty or were postpubertal during the study. More studies are needed on pubertal subjects before conclusions can be drawn. Also, the long term effects of the calcium supplements on the prepubertal subjects are not known.

Another recent study of adolescent females with varying calcium intakes (260 vs. 1637 mg/day) concluded that the main determinant of calcium balance was calcium intake. As calcium intake increased, net calcium absorption increased and urinary calcium did not change. However, this study also concluded that peak bone size, bone mass, and bone mineral content in young women are strongly influenced by genetic information from both mothers and fathers.[67] In the words of the Subcommittee on the Tenth Edition of the RDA's:

> ...the most promising nutritional approach to reduce the risk of osteoporosis in later life is to ensure a calcium intake that allows the development of each individual's genetically programmed peak bone mass during the formative years. The importance of meeting recommended allowance at all ages is stressed, but with special attention to intakes throughout childhood to age 25 years.[8]

C. DISORDERS OF CALCIUM HOMEOSTASIS BEYOND INFANCY

The extracellular ionized calcium concentration must be maintained within narrow limits for its vital and ubiquitous role in normal metabolism. Cardiac contraction is enhanced by increased intracellular Ca^{2+} concentrations and impaired by Ca^{2+} lowering factors. In skeletal muscle, the force of contraction is determined primarily by repetitive Ca^{2+} facilitated release of acetylcholine from the neuromuscular junction. Several enzyme systems within the central nervous system (CNS) (e.g. adenyl cyclase, phosphodiesterase, protein kinase, and others) appear to be regulated by the interaction of Ca^{2+} with calmodulin, a Ca^{2+}-binding protein. This Ca^{2+} calmodulin complex also stimulates myosin kinase in vascular smooth muscle, so that Ca^{2+} influx (enhanced by alpha-adrenergic and inhibited by beta-adrenergic stimuli) causes vasoconstriction. In addition to its vital role in neuromuscular and neurologic function, the concentration of Ca^{2+} plays a critical role in the clotting system and other membrane transport systems. Not surprisingly, alterations in calcium homeostasis frequently have important consequences.

1. Hypocalcemia and Calcium Deficiency

Low calcium states are often, but not always, manifest by abnormally low serum calcium concentrations. Disorders causing total body calcium deficiency may be accompanied by physiologic compensation (i.e., secondary hyperparathyroidism) sufficient to maintain serum calcium levels within the normal range.

Hypocalcemia occurs in a variety of acquired and hereditary diseases. Symptomatic hypocalcemia may occur when ionized calcium falls below 2.5–3.0 mg/dl (0.6–0.7 mmol/l). Regardless of cause, reduced Ca^{2+} levels inhibit acetylcholine release and impair effective muscular contraction in sensory and motor nerves alike. Consequently, a variety of peripheral and CNS effects including tetany, convulsions (grand mal, petit mal, or focal), carpopedal spasms, muscle cramps, twitching, paresthesias, laryngeal stridor, and apnea in the newborn are characteristic features of hypocalcemia. Latent tetany is manifested by Chvostek's (provoked facial muscle twitching) and Trousseau's (provoked carpopedal spasm) signs. Cardiovascular manifestations include hypotension, myocardial depression,[68] congestive heart failure, and arrhythmias.[69]

Decreased total serum calcium concentration, with a normal ionized fraction, can exist with hypoproteinemia. Alkalosis decreases ionized (but not total) serum calcium concentration and may precipitate tetany. Determination of free ionized Ca^{2+} concentration is diagnostic, although the rate of decline may also contribute to the development of symptoms. Estimations of ionized Ca^{2+}, correcting for protein binding and pH, must be done with caution, with total serum hypocalcemia. EKG prolongation of the Q-T interval is helpful in confirming reduced ionized calcium levels, though the correlation is not strong.[70] The etiologies of hypocalcemia are divided according to deficiencies of the various mediators of normal calcium homeostasis (Table 5).

a. Reduced PTH effect

Reduced PTH effect (hypoparathyroidism) can result from deficient parathyroid hormone secretion due to autoimmune or post-surgical parathyroid gland failure, suppression of PTH release (e.g., by chronic hypomagnesemia, maternal hypercalcemia) or insensitivity to PTH (e.g., pseudohypoparathyroidism, PHP). Hypocalcemia and hyperphosphatemia are the major diagnostic clues, resulting from combined deficient calcium release from the bone, decreased $1,25(OH)_2D$ production with resulting decreased intestinal calcium absorption, and decreased renal phosphate excretion. Measurement of serum PTH concentrations distinguishes hormonal deficiency from resistance to hormone action (PHP).

Combined hypocalcemia and hyperphosphatemia resulting from decreased PTH effect lead to physical exam findings independent of the specific cause. Somatic changes include dry, coarse skin, eczematous dermatitis, brittle hair with areas of alopecia, and brittle nails with smooth transverse grooves. Dental abnormalities occur frequently and can be prominent during childhood.[71] These include enamel hypoplasia (circumferential bands or pits traversing the crowns of affected teeth), dental hypoplasia, and failure of adult tooth eruption. Calcification of the lens can lead to cataract formation, which is not reversible by

Table 5 **Etiology of hypocalcemia**

Reduced PTH Effect	**Changes in Ca^{+} Binding or Chelation**
Parathyroid Gland Failure	Alkalosis
Hypoparathyroidism — idiopathic or autoimmune	Respiratory alkalosis
Trauma	Bicarbonate infusion
Postsurgery	Hyperphosphatemia
Post–^{131}I therapy	Renal failure
Infarction	Phosphate administration (e.g., high phosphate formulas, enemas)
Infiltration (e.g., sarcoid, hemosiderosis)	Chemotherapy
Neonatal, resulting from maternal hypercalcemia	
Insensitivity to PTH	Rhabdomyolysis
Pseudohypoparathyroidism	Malignancy
Hypomagnesemia	Pancreatitis
Burns	Fat embolism
Sepsis	Transfusion with citrate-preserved blood
Drugs:	
Aminoglycosides	
Cimetidine	
Cisplatin	
β-adrenergic blockers	
Reduced Vitamin D Effect	**Drugs/Toxins**
Vitamin D Deficiency	Glucagon
Dietary insufficiency	Mithramycin
Increased losses due to:	Calcitonin
Malabsorption	EDTA
Nephrotic syndrome	Protamine
Phenytoin, phenobarbital	Sodium floride
	Colchicine
Impaired Activation of Vitamin D	Theophylline
Renal Disease Vitamin D-dependent rickets (Type I)	Ethylene glycol
Hypoparathyroidism	
Liver failure	
Rhabdomyolysis Resistance to Vitamin D, Vitamin D-dependent rickets (Type II)	

From Allen, D. B., *Pediatric Critical Care,* Fuhrman, B. P., and Zimmerman, J. J., Eds., Mosby-Year Book, St. Louis, 1992, p. 793. With permission.

correction of hypocalcemia. Intracranial calcifications, particularly in the basal ganglia, are identified in approximately 20% of patients with hypoparathyroidism. Rarely, a Parkinsonian-like syndrome can develop during childhood. Development of calcified basal ganglia and cataracts appears to reflect duration of metabolic derangement and is more common in congenital than acquired hypoparathyroidism.

Manifestations of chronic hypocalcemia are often overlooked and the diagnosis delayed, not uncommonly, for years. Scholastic difficulties, impairment of speech and coordination, unexplained episodes of cramping, involuntary contraction of the hands and feet, poor growth, or subjective symptoms such as vaguely characterized paresthesias may be presenting features. Papilledema, elevated cerebrospinal fluid pressure, and neurologic signs mimicking a cerebral tumor should also be considered potential indications of unrecognized hypocalcemia.

Deficient PTH secretion — identified during the first few years of life usually reflects congenital absence or hypoplasia of the parathyroid glands, occurring sporadically or as a component of the DiGeorge syndrome (see Chapter 4). Occasionally, infantile hypocalcemia results from familial hypoparathyroidism, usually inherited as an X-linked recessive trait and, therefore, predominantly affecting males in the first months of life (see Chapter 4). Autosomal dominant and autosomal recessive modes of inheritance have also been observed. Rare associations with ring 16 or 18p or 22p syndromes have been reported. Acquired deficiencies in PTH secretion may follow pathological or iatrogenic parathyroid gland destruction.

Autoimmune-mediated hypoparathyroidism — is suggested by the finding of antibodies directed against parathyroid tissue (identifiable in approximately one-third of young adult affected patients).[72] Anti-endothelial cell antibodies found in patients with idiopathic hypoparathyroidism also implicate antibodies as the cause of endocrine dysfunction.[73] Chronic mucocutaneous candidiasis (resulting from defects in cellular immunity), alopecia areata, and vitiligo (with antibodies directed against melanocytes) may accompany idiopathic hypoparathyroidism and strengthen the hypothesis that immunologic disorders are responsible for deficient PTH secretion. Such patients should be monitored for the possible development of other autoimmune endocrinopathies, particularly primary adrenal insufficiency and, more rarely, primary hypothyroidism, gonadal failure, diabetes mellitus, or pernicious anemia.[74]

Iatrogenic parathyroid gland destruction — is unusual during childhood, due to the relative rarity of hyperthyroidism or hyperparathyroidism requiring surgical intervention. Wide variations in the reported incidence in post-thyroidectomy permanent hypoparathyroidism (0.2–33%) can be attributed to differences in the extent of thyroid resection, surgical expertise, and diligence in diagnosing PTH deficiency.[75] Preferences among pediatric endocrinologists for medical or radioactive iodine treatment for hyperthyroidism has also reduced the occurrence of this complication. Although reported in patients treated for Grave's disease, hypoparathyroidism (either permanent or transient) is an extremely rare complication of radioactive iodine therapy.[76]

Hypocalcemia associated with isolated chronic hypomagnesemia can often be corrected by magnesium replacement, suggesting a role for magnesium in

PTH secretion or action (see Section II.B. for disorders of magnesium metabolism).

Insensitivity to PTH — or pseudohypoparathyroidism (PHP), results in serum biochemical features of hypoparathyroidism (hypocalcemia and hyperphosphatemia) and, usually, elevated levels of parathyroid hormone due to chronic hypocalcemia. The response of target cells to PTH depends on the integrity of a three-component receptor in the cell membrane. The receptor has an extracellular binding site and a guanine nucleotide responsive regulatory G protein. The third component activates adenyl cyclase to synthesize cyclic AMP (cAMP), the second messenger which signals PTH receptor binding to the functional elements within the cell. Target organ unresponsiveness to PTH might be secondary to PTH antagonists, altered PTH receptor binding sites, abnormal G protein, or defective response to cAMP.

In most cases, PHP is inherited as an X-linked dominant trait with variable penetrance, though an autosomal dominant form has been described.[77] Subtypes of PHP are differentiated by their distinct physiologic mechanisms. In type Ia PHP, patients fail to show increased urinary cAMP excretion after intravenous infusion of PTH, indicating a proximal renal tubular defect in hormone action. [78] Reduced activity of G protein, which stimulates adenyl cyclase, appears to account for the limited intracellular cAMP production in response to PTH. There is a 40–50% reduction in the G protein of erythrocytes, platelets, and cultured fibroblasts. Virtually all PHP patients with deficient G protein activity display associated dysmorphic features of Albright hereditary osteodystrophy (AHO), which include short stature, round face, obesity, mental retardation of varying degrees, subcutaneous calcifications or ossification, and bone anomalies, the most distinctive being metacarpal and metatarsal shortening (Figure 8).[78,79] The end-organ resistance to PTH appears to be progressive over the first 2–3 years of life, accounting for the infrequent diagnosis during infancy. Dysmorphic features and possible migratory calcifications may precede hypocalcemia and elevated serum PTH concentrations.

Since deficient G protein activity and, therefore, limited cyclic AMP production in response to hormone stimulation occurs in multiple target cell types, subjects with type Ia PHP have a high incidence of resistance to additional hormones.[80] Primary hypothyroidism, manifest by elevated thyroid-stimulating hormone (TSH) levels, is present in most patients. The absence of antithyroid antibodies and goiter suggest resistance to TSH as the likely mechanism for hypothyroidism.

In PHP type IB, the G protein content of the cells is normal and patients are generally normal in appearance, although a few show typical features of AHO. The exact molecular basis for this type is not well established. However, an intrinsic abnormality in the PTH receptor is suggested by selective resistance to PTH (reduced cAMP response) in skin fibroblasts.[81]

In PHP type II, the typical biochemical features of PHP (hypocalcemia, hyperphosphatemia, elevated serum PTH level) are accompanied by normal urinary cyclic AMP levels but a phosphaturic response to PTH which is

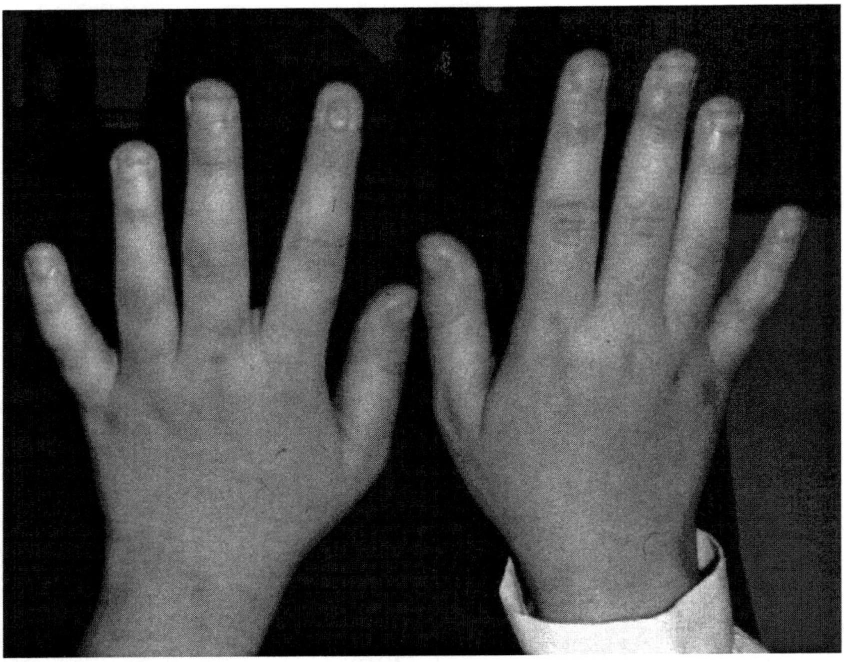

Figure 8 Hands of a child with pseudohypoparathyroidism. Note shortened fourth and fifth fingers (due to metacarpal shortening) and dimpled skin over knuckles.

markedly decreased.[82] End-organ resistance in these patients appears to be due to defective tubular response to cyclic AMP. Because hypocalcemia itself blunts the phosphaturic response to PTH, restoration of normocalcemia should precede definitive testing for type II PHP. For example, patients with hypocalcemia due to vitamin D deficiency have a reduced phosphaturic response to PTH which is restored with establishment of normocalcemia.[83]

The spectrum of specific end-organ resistance to PTH varies in patients with PHP. Direct evidence for skeletal resistance to PTH is lacking. The reduced renal production of $1,25(OH)_2D$ in response to PTH in PHP may be the major factor impairing the normal effect of PTH on bone.[84] In some patients with PHP type I, usually those without physical features of AHO, end-organ resistance to PTH is found in the kidney but bone is normally responsive. High serum concentrations of PTH are accompanied by radiographic evidence of excessive PTH action on bone (osteitis fibrosa cystica), a clinical picture termed pseudohypohyperparathyroidism.[85] Nevertheless, hypocalcemia persists and the basis for this apparent selective renal defect is unknown. On the other hand, patients with pseudopseudohypoparathyroidism (PPHP) demonstrate phenotypic features of AHO, indicating possible skeletal resistance to PTH, but have relatively normal renal cAMP response to PTH and normal serum levels of

Table 6 **Features distinguishing pseudohypoparathyroidism from idiopathic hypoparathyroidism**

	Pseudohypoparathyroidism		Idiopathic Hypoparathyroidism
Age of Onset	Average 8.5 years (98% before 20 years)		Average 16 years (70% before 20 years)
Intelligence	Mental retardation in >60%		Normal in >80%
Female:Male	2:1		1:1
Associated Physical Findings	Short stature Round face Short metacarpals Subcutaneous calcification	(50–75%)	Chronic mucocutaneous candidiasis (16%) Papilledema (18%)

Data from Reference 78.

calcium and phosphorus.[82,85] Interestingly, G protein activity is reduced in patients with PPHP, and other indications of hormonal resistance (e.g., increased TSH levels) may be present.[80] These features are most commonly seen in mothers of patients with classical PHP, suggesting the phenomenon of genetic "anticipation", where inherited defects appear to be amplified in successive generations. Understanding of this disorder remains incomplete.

b. Differential Diagnosis of Hypoparathyroidism, PHP, and PPHP

The clinical features of PHP are those of hypocalcemia and include many of those also found in hypoparathyroidism — tetany, convulsions, carpopedal spasm, muscle cramps and twitching, paresthesias, and laryngeal stridor. However, physical findings differentiate PHP from idiopathic or autoimmune hypoparathyroidism (Table 6).[77] Children with PHP type Ia are of reduced height and thick set, with round faces and short, thick necks. The digits of the hands and feet are short and stubby, and because of shortened metacarpals, the skin over the knuckles is dimpled. Classically, the index finger is longer than the middle finger because the second metacarpal is not affected by metacarpal shortening (Figure 8). These features are not found in patients with hypoparathyroidism. Subcutaneous tissue calcifications and mental retardation are also typical for PHP.

Radiological features also aid in distinguishing PHP from hypoparathyroidism. Subcutaneous bone formation appears as punctate or granular lesions, especially in the extremities. Shortened metatarsal and metacarpal bones, with sparing of the second digit, are characteristic of PHP. The bones have widened medullary canals and coarse trabecular patterns. Thickening of the calvarium also suggests PHP. In some patients with PHP type I (osteitis fibrosa cystica), characteristic hyperparathyroid bone resorptive lesions are present. On the contrary, hypoparathyroidism has few distinctive radiologic findings, limited primarily to osteomalacia secondary to vitamin D deficiency in prolonged unrecognized cases.

Table 7 **Biochemical characteristics in types of hypoparathyroidism**

Characteristic	Serum PTH	U_{cAMP}	Exogenous PTH[a] U_{cAMP} Response	Exogenous PTH[a] U_p Response
Deficient PTH section	Low	Low	Normal	Normal
Secretion of ineffective PTH	High[b]	Low	Normal	Normal
Resistance to PTH-defective receptor-adenylate cyclase complex	High	Low	Low	Low
Resistance to PTH-defect distal to adenylate cyclase complex	High	High	Normal	Low

[a] 5 U/kg body weight, 200 U maximum, of synthetic (1-34) NH2-terminal fragment of PTH administered intravenously over 10 minutes. Urine collected before and after infusion for measurement of cAMP (U_{cAMP}) and phosphorus (U_p).
[b] High if species secreted cross-reacts with antiserum used in radioimmunoassay.

From Aurbach, G. D., Marx, S. J., and Spiegel, A. M., *Williams Textbook of Endocrinology*, Wilson, J. D. and Foster, D. W., Eds., W.B. Saunders, Philadelphia, 1992. With permission.

Hypocalcemia, hyperphosphatemia, and normal renal function strongly suggest deficient PTH secretion or effect. In addition to clinical features described above, definitive diagnosis requires determination of plasma PTH concentrations, which are high in PHP but low in hypoparathyroidism. Measurement of urinary cAMP levels and the response in urinary cAMP and phosphate excretion to exogenous PTH (or synthetic NH_2-terminal [1-34] fragment) allows further determination of the presence and site(s) of PTH resistance (Table 7).[86] Patients with deficient PTH secretion respond normally to PTH infusion. In PHP type I, urinary cyclic AMP and phosphorus are not increased following PTH administration. In type II PHP, urinary cyclic AMP is increased but the hyperphosphaturic response does not occur. In PHP type I with osteitis fibrosa cystica, elevated serum PTH levels produce the picture of hyperparathyroid bone disease despite hypocalcemia. In PPHP, phenotypic features of PHP are present but intact bone and renal responsiveness to PTH result in normal serum calcium and phosphorus concentrations.[87]

c. Treatment of Hypocalcemia

Goals of treating hypocalcemia during childhood include: 1) normalization of serum calcium and the alleviation of neuromuscular symptoms; 2) preventing vitamin D intoxication and its accompanying hypercalciuria, hypercalcemia, and risk for nephrocalcinosis and cataracts; and 3) promoting normal growth and development of the skeleton. Treatment of hypocalcemia, regardless of etiology, should be preceded by consideration of readily treated or confounding factors (e.g., acute respiratory alkalosis). Rapid development of hyperphosphatemia suggests acute renal failure, cell lysis, or excessive supply.

Efforts should be made to reduce serum phosphate levels, since intravenous calcium therapy may cause metastatic deposition of calcium-phosphate salts. Chronic hypomagnesemia impairs both release of and response to PTH, and thus correction of hypocalcemia will require correction of hypomagnesemia.[88] Critically ill patients may develop hypomagnesemia because of gastrointestinal or renal losses of Mg^{2+} or inadequate intake.[89]

Urgency of therapy is determined by the child's clinical status. Asymptomatic hypocalcemia is appropriately treated with oral calcium salts. For the seriously ill patient with evolving hypocalcemia, replacement therapy is accomplished with infusion of intravenous calcium gluconate (10% solution: 90 mg elemental calcium per 10 ml) at 75 mg of elemental calcium/kg/day. For impending cardiovascular or neuromuscular collapse, when a prompt and predictable rise in Ca^{2+} is required, 0.1 ml of $CaCl_2$ per kilogram or 2 ml/kg of body weight of 10% calcium gluconate given over 5–10 minutes is generally sufficient. The possibility of bradycardia and asystole with infusion of Ca should be anticipated, with cardiac monitoring and atropine at hand. Care is required to avoid tissue damage by extravasation, precipitation with concomitantly-administered bicarbonate, and untoward cardiac rhythm disturbances in digitalized patients. In appropriate cases, prompt supplementation with magnesium or vitamin D (for vitamin D deficiency or lack of PTH effect) will be needed for maintenance of normocalcemia.

When the diagnosis of either deficient PTH secretion or end-organ unresponsiveness to PTH is confirmed, vitamin D therapy is required to restore serum $1,25(OH)_2D$ concentrations and optimize intestinal calcium absorption. This can effectively be done with $1,25(OH)_2D$ (0.03–0.08 μg/kg/day up to a maximum of 1–2 μg/day) or dihydrotachysterol (23,25-dihydroxyvitamin D, 250-1000 μg/day), which has biologic activity midway between 25OHD and $1,25(OH)_2D$. Treatment should begin with low doses, with gradual increases and frequent monitoring of serum calcium concentrations until normal levels are achieved. To avoid hypercalciuria and nephrolithiasis, long-term therapy should maintain serum calcium concentrations in the range of 8–9 mg/dl (2–2.2 mmol/l). Administration of oral calcium salts is often required; calcium carbonate, which is 40% calcium and therefore provides 1.0 g of elemental calcium in 2.5 g calcium carbonate, is the most practical oral preparation. Restoration of serum calcium levels decreases the renal threshold of phosphorus excretion and lowers serum phosphorus concentrations. Nevertheless, antacid therapy directed at reduction of serum phosphorus is often required during early childhood, particularly when reliance on dairy products for caloric intake is high or dietary compliance is low. Calcium acetate salts (250–750 mg/day) now provide an attractive alternative to aluminum hydroxide, an antacid which has been associated with development of aluminum intoxication.[90]

Periodic monitoring of serum calcium and phosphorus levels and urinary calcium excretion aid in avoiding vitamin D overdosage, hypercalciuria and nephrolithiasis, and extra-osseous calcification. The overall prognosis for hypoparathyroidism and pseudohypoparathyroidism is good, particularly if diagnosis and treatment are begun early in the congenital forms.

2. Reduced Vitamin D Effect

Reduced vitamin D effect results from nutritional vitamin D deficiency (e.g., fat malabsorption), impaired activation of vitamin D (e.g., renal disease-associated or genetic deficiency of 25OHD 1-α-hydroxylase), or exaggerated losses or metabolism (e.g., phenytoin therapy). Rickets resulting from primary calcium deficiency is extremely rare[91] and rickets in infants secondary to pure vitamin D deficiency with hypocalcemia, which almost always result from a combination of prolonged breast feeding and limited exposure to sunlight, is discussed elsewhere (see Chapter 4). Nutritional vitamin D deficiency beyond infancy is exceedingly rare in the absence of complicating factors (e.g., malabsorption or liver disease). Therefore, this discussion will address disorders of vitamin D metabolism (vitamin D-dependent rickets [VDDR] type I and renal osteodystrophy), vitamin D resistance (VDDR type II), and rickets secondary to renal phosphate wasting (hypophosphatemic [XLH] rickets).

Insufficient vitamin D action from any cause leads to deficient calcium and phosphorus absorption, poor bone mineralization, and the characteristic bone deformities of rickets in growing bone. Deficiencies in bone mineralization are most evident in regions of rapid bone growth and turnover (e.g., cranium, wrists, and ribs). However, after the first year of life, skeletal deformities are most evident in the lower extremities due to weight bearing. Early widening of lower extremity growth plates and metaphyseal irregularities are accompanied by genu valgum (knock-knee) or genu varum (bow-leg), which worsens progressively in inadequately treated cases. Treatment before age 4 years may lead to resolution of moderate deformities, but persistent disease beyond early childhood leads to permanent deformity, adult short stature, and susceptibility to pathologic fractures. Dental development is delayed and erupted teeth show irregular pits and enamel hypoplasia. Proximal muscle weakness, prominent in older children, can compromise respiratory function in severe cases.

Rickets is caused by deficient mineralization of growing bone. Failure to calcify cartilage is reflected radiologically by delayed opacification of the epiphyses and widening of the growth plates (Figure 9). The ends of metaphyses are irregular and become widened due to pressure on poorly mineralized bone. Shaft deformities are usually obvious, but pathologic fractures are rare during childhood. In certain disorders where calcium malabsorption is prominent, variable signs of secondary hyperparathyroidism can be seen. Metaphyseal growth arrest lines, parallel to the growth plates, may indicate intervals of variable disease severity.

a. Impaired Vitamin D Metabolism or Action

Vitamin D-dependent rickets type I (VDDR type I) is caused by an absent or inactive form of renal 25OHD-1-α-hydroxylase, with the result that serum

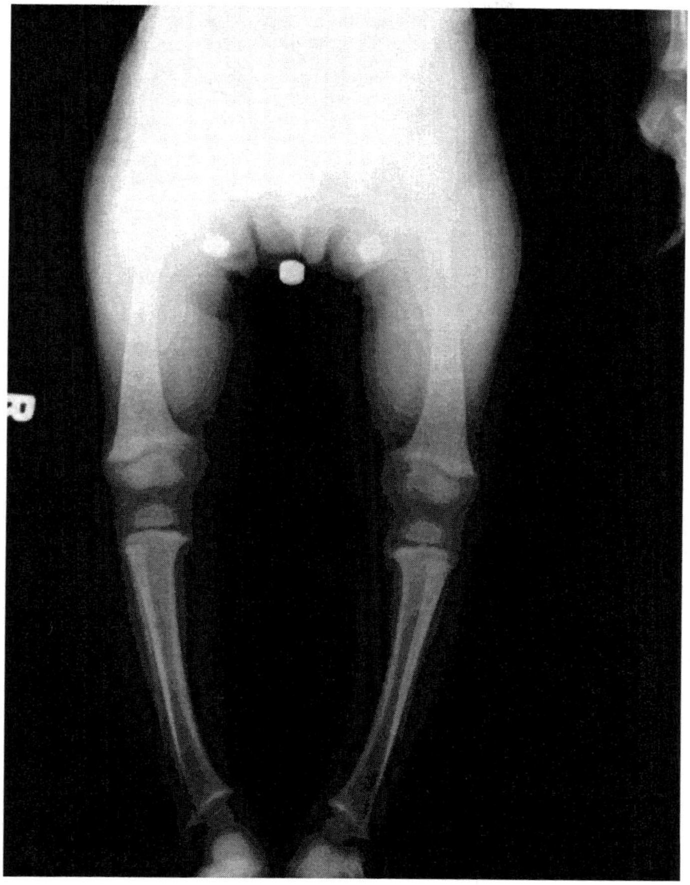

Figure 9 Classic vitamin D deficient rickets. Note bowing of legs and distal femoral metaphyseal flaring.

$1{,}25(OH)_2D$ is severely reduced.[92] This rare autosomal recessive disorder leads to development of characteristic clinical and biochemical features, which mirror nutritional vitamin D deficiency, usually during the first year of life.[93] (For a discussion of this disorder see Chapter 4).

Compromise of renal 25-OHD-1-α-hydroxylase activity occurs most commonly as a component of chronic renal failure. In addition to reduced vitamin D activation secondary to limited functional renal mass, phosphate retention further diminishes the activity of the 25-OHD-1-α-hydroxylase system. These factors, combined with an independent inhibition by azotemia of intestinal calcium absorption and lower serum calcium levels, promote the development of secondary hyperparathyroidism (Figure 10). Bone mineral loss is accelerated by chronic metabolic acidosis. Aluminum intoxication from aluminum contamination of dialysis fluids may further compromise skeletal integrity. The resulting bone disease, collectively termed renal osteodystrophy, may

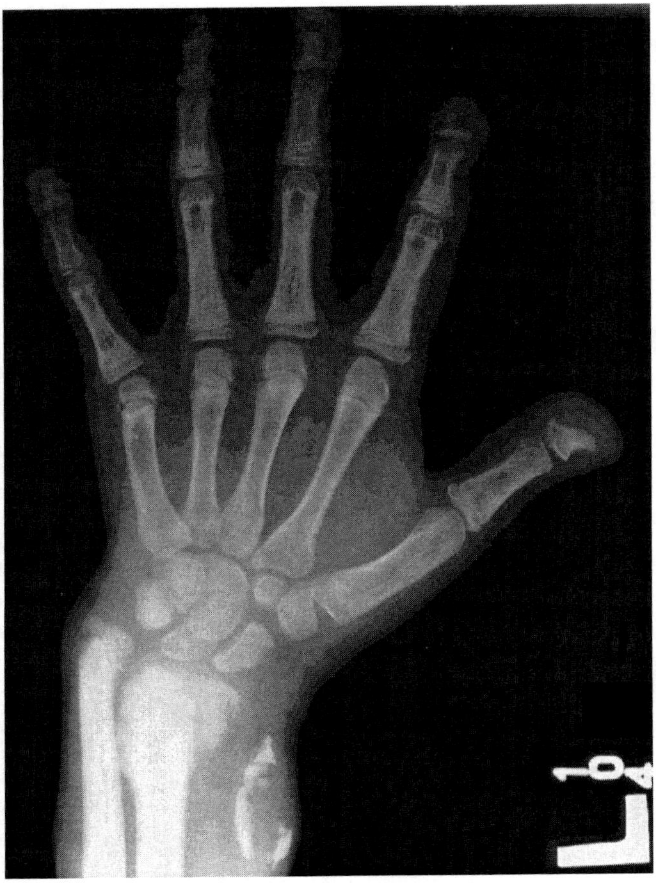

Figure 10 Hyperparathyroidism secondary to chronic renal insufficiency. Note subperiosteal resorption of radial edge of phalanges and of phalangeal tufts.

include osteopenia, pathologic fractures, and ectopic calcifications. Treatment of secondary hyperparathyroidism and renal osteodystrophy includes vitamin D or its metabolites, phosphate binders, dietary phosphate restriction and, in particularly severe cases, partial parathyroidectomy.

b. Resistance to Effect of $1{,}25(OH)_2D$

Vitamin D dependent rickets type II (VDDR type II), often called hypocalcemic vitamin D-resistant rickets, is caused by point mutations affecting the structure or function of the receptor for $1{,}25(OH)_2D$. Renal production of $1{,}25(OH)_2D$ is normal or even supranormal. There is a great heterogeneity in the clinical features which include variable degrees of rickets, hypocalcemia, and secondary hyperparathyroidism.[94] Treatment is very difficult because of the resistance to all vitamin D metabolites. A recent form of therapy which shows promise is long-term parenteral calcium and phosphorus supplementation.[95]

Treatment with anticonvulsant medications may also interfere with vitamin D metabolism and action to an extent sufficient to cause hypocalcemia, rickets, or osteomalacia in patients with epilepsy.[96] Risk for this adverse effect is compounded by multiple anticonvulsant therapies. Anticonvulsant drugs stimulate hepatic microsomal oxidase activity, and convert vitamin D and 25OHD to more polar, biologically inactive metabolites. A decline in serum 25OHD concentration is frequently accompanied by secondary hyperparathyroidism.[97] Serum concentrations of $1,25(OH)_2D$, while usually normal, may be inappropriately low for the degree of secondary hyperparathyroidism. Even more important effects of anticonvulsants appear to be the inhibition of intestinal calcium absorption and calcium mobilization from bone (interference with peripheral actions of vitamin D).[98,99] Calciferol analogues can be used to prevent or treat this complication.

3. Alterations in Phosphate Metabolism

Alterations in phosphate metabolism are important contributors to the development of rickets and osteomalacia, and may cause hypocalcemia. Because phosphate is ubiquitously distributed in all human foodstuffs, nutritional phosphate deficiency does not ordinarily occur. However, phosphate depletion may develop during treatment with phosphate chelators, intravenous feeding or hemodialysis using a low-phosphate solution. Such depletion is normally compensated for by nearly complete cessation of renal phosphate excretion. Renal 25OHD-1-α-hydroxylase activity is also increased, promoting intestinal phosphate (and calcium) absorption. However, rickets will develop if sufficient phosphate is not available to support mineralization of growing bone.

Acute changes in Ca^{2+}-binding equilibrium or chelation are perhaps the most frequent causes of hypocalcemia in the ill child. Hyperphosphatemia or respiratory alkalosis may dramatically reduce total and ionized calcium. Infusion of large amounts of citrate-preserved blood and acute phosphorus overload or retention (e.g., acute renal failure) rapidly deplete extra cellular fluid calcium concentrations. Various drugs (see Table 5) may also contribute to the development of hypocalcemia.

a. Hypophosphatemic Rickets

Hypophosphatemic rickets (XLH) is a familial disorder with renal phosphate wasting and hypophosphatemia. The most common cause of rickets, it was the first to be recognized as vitamin D resistant.[93] This disease is most often transmitted as an X-linked dominant trait; however, genetic heterogeneity exists, with kindreds demonstrating autosomal dominant or recessive transmission patterns. About one-third of patients present with a negative family history and are presumed spontaneous mutations. The mutant gene has been mapped on the short arm of the X chromosome and the number of sporadic cases indicates that the mutation rate at this locus must be fairly high.[100]

The primary defect in XLH rickets is isolated impairment of phosphate transport in the proximal renal tubule. In untreated patients, hypocalcemia is usually not present, serum PTH concentrations are normal, and the renal wasting of phosphate is not affected by exogenous PTH. Thus, the XLH rickets mutation seems to block the component of phosphorus transport in the kidney normally responsive to PTH. However, the precise target tissue defect has not been defined. A parallel defect in intestinal absorption of phosphate may be present. Mildly decreased intestinal absorption of calcium and $1,25(OH)_2D$ concentrations, which are only slightly higher than age-matched controls in untreated XLH rickets patients, suggests that a defective or inappropriate response of 25-OHD-1-α-hydroxylase to either PTH or phosphate depletion may occur in this disorder.[100] However, the long-term administration of $1,25(OH)_2D$ to these patients does not improve hypophosphatemia or promote the healing of rickets.[100]

Fasting hypophosphatemia is the hallmark of this disorder and is detectable from age 6 months to old age.[101] Hemizygote males, who often manifest florid rickets during childhood, are more severely affected than heterozygotic females, suggesting a gene-dose effect. Clinical features of this disease include progressive lower limb deformities and stunted growth. Although hypophosphatemia is present in infancy, the general well-being of the patient obscures the diagnosis until leg deformities develop with the onset of weight bearing. Radiologic examination reveals classic findings of rickets. Dental development is delayed, but teeth have normal enamel, in contrast to the dental enamel hypoplasia of children suffering from hypocalcemic types of rickets. Sensorineural hearing defects occur with increased frequency in XLH rickets, which may be due to cochlear defects linked to a separate X chromosome locus.

Because the renal defect in XLH rickets is associated with both phosphate wasting and defective 25-OHD-1-α-hydroxylase, treatment for this disorder combines $1,25(OH)_2D$ and phosphate salts. Administration of pharmacologic doses of calciferol alone may improve skeletal defects, but will not correct hypophosphatemia and may result in vitamin D intoxication. $1,25(OH)_2D$ improves intestinal phosphate absorption, reduces the amount of daily phosphate supplementation needed, and alleviates secondary hyperparathyroidism associated with phosphate therapy. Linear growth and bone mineralization are significantly improved with this combined treatment regimen. If begun before 6 years of age, permanent bone deformities may be prevented or kept to a minimum. Patients require frequent clinical monitoring to avoid renal damage from hyperphosphaturia, hypercalciuria, and nephrocalcinosis.[100]

A variant form of this type of rickets has recently been described and is called hereditary hypophosphatemic rickets with hypercalciuria. It is distinguished from XLH rickets by the presence of muscle weakness and high serum $1,25(OH)_2D$ concentrations.[100]

4. Hypercalcemia

a. Clinical and Laboratory Diagnosis

In contrast to dramatic neuromuscular manifestations indicative of hypocalcemia, the physical effects of hypercalcemia may be subtle and nonspecific. These include general malaise, failure to thrive, polyuria and thirst, and abdominal pain. Physiologic effects on the renal, cardiovascular, and central nervous systems predominate and reflect both the degree of calcium elevation and its duration. The increased filtered load of calcium creates hypercalciuria and accompanying polyuria, reduced urine concentrating ability, dehydration, and eventual kidney stone formation. Hypertension is common, possibly through increased renin production or direct effects on peripheral vasoconstriction and inotropism.[102] Alterations in the cardiac conduction system include a shortened Q-T interval and a tendency to arrhythmias. Impaired nerve conduction resulting from excess calcium creates hypotonia, hyporeflexia, and even paresis in severe cases. Alterations in central nervous system function, including lethargy, confusion, hallucinations, and even coma can occur. Constipation, anorexia, and abdominal pain are frequent complaints, and result from reduced intestinal motility and, occasionally, pancreatitis. Promotion of gastrin release by calcium may account for an increased incidence of peptic ulcer disease. Soft tissue deposition of calcium can impair function of the lungs, kidneys, blood vessels, and joints. A serum calcium concentration in excess of 15 mg/dl (3.75 mmol/l) represents a medical emergency.

In the absence of hyperproteinemia, determination of elevated serum Ca levels reliably indicates increased Ca^{2+} concentrations. Because common causes of hypercalcemia in adults (hyperparathyroidism and malignancies) are less common in children, hypercalcemia occurs less frequently than hypocalcemia during childhood. Diagnostic possibilities are best approached by considering the predominant underlying mechanism of hypercalcemia. Measurement of a logical combination of serum calcium, phosphorus, vitamin D metabolites, and immunoreactive PTH, and urine calcium and phosphorus excretion will, with rare exceptions, reveal the underlying disorder (Table 8). Increased bone resorption occurs with excess PTH effect (e.g., hyperparathyroidism, ectopic PTH), immobilization, or bone lysis by metastatic malignancy (e.g., neuroblastoma).

b. Increased Bone Resorption

Hyperparathyroidism — (PTH-mediated hypercalcemia) is distinguished by a depressed serum phosphate concentration, decreased renal tubular reabsorption of phosphate, and a PTH level inappropriately elevated for the simultaneous serum Ca^{2+} level. Hypercalcemia results from PTH-mediated increased bone resorption, relatively impaired urinary calcium excretion (though hypercalciuria is present), and increased intestinal calcium absorption (due to enhanced $1,25(OH)_2D$ synthesis).[103]

Primary hyperparathyroidism is rare during childhood. Recognition may increase with greater reliance on "routine" chemistry profiles in pediatrics. Sporadic neoplastic transformation of parathyroid glands is invariably benign (parathyroid adenoma) and occurs most often in males between the ages of 6–15 years. Ultrasound and/or radionucleotide imaging are useful aids in preoperative diagnosis and localization.[104] A large proportion of cases,

Table 8 **Hypercalcemia**

Predominant Mechanism of Hypercalcemia	Disorder	Serum					Urine	
		Ca	PO_4	25(OH)–Vitamin D	$1,25(OH)_2$–Vitamin D	iPTH	Ca	$TmPO_4/GFR$
Increased bone resorption	Hyperparathyroidism	H	L	N	H	H	H	L
	Malignancy-ectopic PTH	H	L	N	N or L	L	H	L
	Immobilization, thyrotoxicosis, or bone lysis	H	N or H	N	L	L	H	H
Increased intestinal calcium absorption	Vitamin D intoxication	H	N or H	H	N or H	L	H	H
	Sarcoidosis	H	N or H	N	H	L	H	H
	? Idiopathic infantile hypercalcemia or Williams syndrome	H	N	N	L, N or H	L	H	N or H
	Abrupt glucocorticoid withdrawal or deficiency	H	N or H	N	N	L	H	N or H
Decreased renal excretion of calcium	Familial hypocalciuric hypercalcemia	H	N or L	N	N or H	N or H	L	L

H, high; L, low; N, normal.

particularly during infancy, are familial and due to parathyroid gland hyperplasia. Inheritance of familial hyperparathyroidism is most often autosomal dominant, (though recessive inheritance has been described)[105-108] and is genetically related to multiple endocrine neoplasia types I (associated with pituitary and pancreatic tumors) and IIa (associated with medullary carcinoma of the thyroid and pheochromocytoma). Severe neonatal hyperparathyroidism may occur in infants with familial hypocalciuric hypercalcemia.[109]

Many pediatric patients with primary hyperparathyroidism have a serum calcium greater than 15 mg/dl (3.75 mmol/l). Constitutional symptoms of hyperparathyroidism (e.g., abdominal pain, constipation, fatigue, weakness) reflect the severity of hypercalcemia. In addition, osteolytic activity of PTH may produce bone pain, and long-standing hypercalciuria is associated with development of renal colic, renal stones, and hypertension. Radiographic findings include generalized bone demineralization, subperiosteal resorption, lytic lesions of the skull, cysts in the pelvic and long bones, and chondrocalcinosis.

Following the initial medical management for symptomatic hypercalcemia (detailed below), optimal treatment for primary hyperparathyroidism is surgical excision of hyperplastic or neoplastic parathyroid tissue. Relatively poor outcomes have been observed in patients with infantile diffuse parathyroid gland hyperplasia, with mortality up to 87.5% with medical management alone and 24% with surgical intervention.[110] Total parathyroidectomy with oral replacement of calcium and vitamin D or autotransplantation of parathyroid tissue has been advocated.[111,112]

Secondary hyperparathyroidism — occurs in response to chronic hypocalcemia or calcium deficiency. As discussed elsewhere (see Chapter 4), maternal hypoparathyroidism and vitamin D deficient rickets or hepatic rickets may cause secondary hyperparathyroidism in the fetus and newborn. With prolonged compensatory PTH hypersecretion, radiographic findings are similar to those of primary hyperparathyroidism. In patients with chronic renal insufficiency, the "set point" of serum calcium at which parathyroid hormone secretion is suppressed is increased.[113] Hypercalcemia may occur in these patients and may be aggravated by concomitant vitamin D therapy. Renal osteodystrophy secondary to renal failure results from the combined effects of disordered vitamin D metabolism and secondary hyperparathyroidism.

Neoplastic disease — causes hypercalcemia less frequently in children than adults, but may do so by three mechanisms: 1) diffuse skeletal infiltration by hematologic malignancy; 2) destruction of bone by metastases from solid tumors; and 3) secretion by solid tumors of parathyroid hormone-related peptide (PTHrP). Acute lymphoblastic leukemia is the malignancy most frequently associated with hypercalcemia during childhood.[114] Management of each situation includes treatment of symptomatic hypercalcemia and resection or dissolution of the tumor, if possible. In selected circumstances, corticosteroids, mithramycin, calcitonin, or diphosphonates may provide valuable adjuncts for lowering serum calcium levels.[115]

Immobilization — perhaps, the most common cause of hypercalcemia due to increased bone resorption during childhood. Typically, this complication occurs when an active adolescent is immobilized in a cast following trauma or surgery. Children with burns and spinal cord injuries are also at risk. Hypercalciuria and occasionally hypertension usually precede hypercalcemia. Suggested mechanisms include decreased mechanical stress on bone, altered osteoblastic and osteoclastic activity,[116] and increases in metabolic products of vitamin D. Children with hypervitaminosis A or hyperthyroidism may also develop hypercalcemia secondary to a direct stimulatory effect of vitamin A or thyroxine and triiodothyronine on bone resorption.[117,118]

c. Increased Intestinal Calcium Absorption

Increased intestinal calcium absorption with hypercalcemia can be due to vitamin D intoxication, increased sensitivity to vitamin D (proposed for idiopathic hypercalcemia of infancy), or ectopically produced $1,25(OH)_2D$ (sarcoidosis). Serum phosphate concentrations and maximal tubular reabsorption of phosphate are normal or increased, and PTH concentrations are suppressed in these disorders. Detection of an elevated 25OHD concentration (nutritional excess) or $1,25(OH)_2D$ concentration (sarcoidosis, hyperparathyroidism) may be helpful.

Vitamin D intoxication — with hypercalcemia may develop during administration of any exogenous vitamin D preparation at any time during treatment. This occurs most commonly in children receiving $1,25(OH)_2D$ supplementation for renal insufficiency and less frequently in children treated for hypoparathyroidism and PHP. Marked elevation in the appropriate vitamin D metabolite establishes the diagnosis; for example, if vitamin D_2, D_3, or 25-OHD is being taken, high serum 25OHD concentrations (with normal or slightly elevated $1,25(OH)_2D$ concentrations) are found. In addition to enhanced intestinal calcium absorption, hypercalcemia and hypercalciuria are exacerbated by increased osteoclastic bone resorption and diminished PTH-mediated tubular reabsorption of calcium.

Treatment of vitamin D intoxication begins with withdrawal of vitamin D analogues. The expected duration of hypercalcemia following intervention varies with the metabolite ingested; it is more prolonged with vitamins D_2 and D_3 and shorter with the more active metabolites. Presumably, through similarities in steroid structure, glucocorticoids antagonize the stimulatory effect of vitamin D on intestinal calcium absorption and are particularly effective in the treatment of vitamin D-mediated hypercalcemia. Severe hypercalcemia may require additional interventions (e.g., saline diuresis, calcitonin) detailed below.

Another disorder which occurs in childhood associated with hypercalcemia, idiopathic infantile hypercalcemia or Williams Syndrome, may be secondary to endogenous production of or enhanced sensitivity to vitamin D.

Sarcoidosis — increased plasma concentrations as well as sensitivity to vitamin D appear to cause hypercalcemia, hypercalciuria, and enhanced intestinal calcium absorption in children with sarcoidosis.[119,120] Extrarenal conversion of 25OHD to 1,25$(OH)_2$D is suggested by the occurrence of hypercalcemia in an anephric patient with sarcoidosis, and 1-α-hydroxylase activity has been demonstrated in alveolar macrophages and lymph node homogenates obtained from sarcoid patients.[121,122] Gamma-interferon, which can enhance production of 1,25$(OH)_2$D and is increased in alveolar macrophages and lymphocytes in sarcoidosis, may be central to the pathogenesis of this effect.[123] Normal feedback inhibition of 1-α-hydroxylase activity by hypercalcemia or exogenous vitamin D administration is lacking.[124] Nephrolithiasis and renal insufficiency may result from untreated hypercalciuria. Similar explanations have been proposed to explain hypercalcemia associated with tuberculosis, histoplasmosis, silicone-induced granulomas, and other granulomatous disorders.

Glucocorticoid deficiency — resulting from either adrenal insufficiency or withdrawal of exogenous glucocorticoids, is infrequently associated with hypercalcemia. The precise etiology of this effect remains unknown. Unopposed vitamin D stimulation of intestinal calcium absorption may raise serum calcium levels during acute withdrawal of pharmacologic glucocorticoids. Hemoconcentration, enhanced mobilization of skeletal calcium, and increased renal absorption of calcium are more likely contributing factors in patients with Addison's disease.[125]

d. Decreased Renal Excretion of Calcium

Decreased renal excretion of calcium occurs with dehydration or treatment with thiazide diuretics and lithium, and frequently aggravates the severity of hypercalcemia in patients with unrecognized hyperparathyroidism.[126] Familial hypocalciuric hypercalcemia (FHH), an autosomal dominant disorder, is characterized by serum calcium levels 0.5–1.5 mg/dl above the upper limit of the normal range, normal (but not appropriately suppressed) serum PTH concentrations, and inappropriately low rates of urinary calcium excretion. Affected patients are usually asymptomatic, but may have symptoms of mild hypercalcemia. Pancreatitis rarely occurs. Renal function is normal, but calcium excretion is reduced approximately 50% when compared to patients with hyperparathyroidism and similar serum calcium levels, indicating enhanced tubular reabsorption of calcium. The pathogenesis of FHH remains uncertain. Parathyroid gland growth and PTH secretion are not regulated by calcium in the usual manner and the renal tubule, independent of PTH, reabsorbs an excessive fraction of filtered calcium.[127] Hypersensitivity of the cAMP system to PTH has been suggested as a possible underlying defect in FHH, but is contradicted by the lack of effect of total parathyroidectomy on urine calcium excretion and other *in vitro* studies.[127,128] Risk for severe hypercalcemia is greatest during the neonatal period and may be associated with homozygosity for the FHH gene.[129] During childhood and beyond, a benign clinical course is expected, and no treatment is required.

e. Treatment of Hypercalcemia

A serum Ca level above 15 mg/dl (3.75 mmol/l) may be life threatening and requires direct Ca-lowering therapy in addition to attention to the underlying disorder. Extreme hypercalcemia usually indicates reduced glomerular filtration. Hydration with isotonic saline (200-250 ml/kg/day) followed by furosemide diuretics (1 mg/kg every 6 hours) will result in calciuresis and amelioration of hypercalcemia in the vast majority of cases. Excessive losses of sodium, potassium, magnesium, and PO_4 may occur. Thiazide diuretics should *not* be used because of their hypercalcemic effects.

Adjunct therapy is directed at the specific cause of hypercalcemia. Drugs which inhibit excessive bone resorption include calcitonin (4–8 units/kg given subcutaneously or by intramuscular injection every 4–6 hours), mithramycin (25 mg/kg IV for 4 hours), ethane-1-hydroxy-1, 1-diphosphonate (etidronate; 7.5 mg/kg/day IV), and indomethacin (1 mg/kg/day). Calcitonin, now available in either salmon or human (recombinant DNA) form, blocks PTH-induced bone resorption through a direct effect on osteoclasts, facilitates calciuria, is relatively non-toxic, and has a prompt (but transient) peak effect at 1 hour.[130] Mithramycin is a toxic antibiotic that inhibits osteoclastic activity within 24 hours but has potential side-effects including thrombocytopenia, and hepatic and renal toxicity. Intravenous etidronate is effective in the treatment of hypercalcemia of malignancy. Oral administration can be utilized for chronic therapy, but the hypocalcemic effect is delayed and may result in impaired mineralization. Indomethacin or aspirin is useful when excessive PGE_2 production is suspected (e.g., some tumor-related hypercalcemic syndromes). Intestinal calcium absorption is reduced by corticosteroids (hydrocortisone 1 mg/kg every 6 hours or equivalent). Consequently, glucocorticoids are especially useful in treatment of vitamin D-related hypercalcemia. Their delayed onset of action, however, limits their use in the acute setting. Phosphates are not recommended because of the likelihood of soft tissue deposition of calcium-phosphate salts.

II. MAGNESIUM

A. NORMAL MAGNESIUM METABOLISM IN CHILDREN

Magnesium is second in abundance only to potassium as an intracellular cation and is an important regulator of cellular processes. Magnesium is the central part of every molecule of chlorophyll, the ultimate source of all stored energy for life. Necessary for high energy phosphate (ATP formation), it is therefore required for synthesis, growth, thermogenesis, and motility. Magnesium is also a cofactor for more than 300 enzyme systems.

The adult human contains 24 grams (2,000 mEq) of magnesium, of which 60% is located in the skeleton, a pool which is very slow to turn over. Thirty-nine percent is distributed almost equally between muscle and nonmuscular soft tissue in the intracellular space, bound mainly to protein and the energy rich phosphates. Only 1–2% is found extracellularly in a rapid turnover pool, where 30% is protein bound. Most of the remaining 70% in the circulation is free and ionized and available for biochemical processes and filtration by the kidney.

Like calcium, serum concentrations of magnesium are maintained within a narrow range of 1.5–2.0 mg/dl (0.7–1.0 mmol/l). However, unlike calcium, there is no known homeostatic system regulating serum magnesium concentration, which remains remarkably constant among individuals and does not change significantly with age after the first year of life. Since magnesium is predominantly an intracellular cation, its plasma concentration may not accurately indicate total body magnesium content. The concentration of magnesium in erythrocytes is three times higher than in plasma and reflects plasma magnesium concentration at the time of erythropoeisis.[131] However, the correlation of red cell magnesium with total body content is poor. The concentration of magnesium in mononuclear cells has been reported to provide a better estimate of total body magnesium content.[132]

The current recommended daily allowance of magnesium is 6 mg/kg/day for years 1–15 of life. For ages 15–18, 300 mg/day is recommended for females, and 400 mg/day (7–10 mg/kg/day) for males.[8] However, there is little data available upon which to base the RDA for children. Factors influencing pediatric magnesium requirements and metabolism include: 1) genetic and environmental factors, 2) hormonal factors, and 3) factors affecting magnesium availability and conservation, including dietary intake, intestinal absorption, and urinary retention/excretion.

1. Genetic and Environmental Factors

Genetic and environmental factors influencing magnesium requirements and metabolism are not as well defined as those for calcium, though in general, larger individuals have higher total body magnesium content and thus higher absolute requirements. Overall nutrition and health status play major roles, as the deficiency state is rarely seen except in patients with chronic illnesses and in debilitative states.

2. Hormonal Factors

Serum magnesium concentration is maintained largely independent of the known calcitropic hormones. Nevertheless, several hormones exhibit important influences on magnesium metabolism and are also influenced by magnesium. In healthy subjects, PTH is not essential for the regulation of magnesium balance. Large doses of exogenous PTH increase both intestinal absorption and renal reabsorption of Mg, but small doses have no effect. Mild magnesium deficiency increases release of PTH while hypermagnesemia suppresses the synthesis and/or release of the hormone from the parathyroid glands and impairs its peripheral actions on bone, kidney, and intestine.[133,134] However, severe magnesium deficiency also inhibits the release of the hormone from the gland.[135] Impaired PTH release and action during severe hypomagnesemia have both been attributed to defective activation of magnesium-dependent adenyl cyclase.[136]

Vitamin D appears to increase gastrointestinal absorption of magnesium, but these studies are complicated by simultaneous changes in calcium and phosphorus metabolism.[137] Magnesium is required for the hepatic 25-hydroxylation of vitamin D. Not surprisingly, given the influence of magnesium on PTH secretion and vitamin D metabolism, hypomagnesemia is often complicated by hypocalcemia. In these circumstances, correction of magnesium depletion is required before sustained correction of hypocalcemia can be achieved. Thiamin is another vitamin with important interactions with magnesium. When administered to a magnesium-deficient animal, thiamin is not utilized and may accentuate symptoms of magnesium depletion.[138]

Insulin causes a shift of magnesium from the extracellular to the intracellular space, thus conserving total body content of magnesium while lowering serum magnesium concentrations. Renal and gastrointestinal absorption of magnesium are not directly augmented by insulin administration, but markedly impaired renal conservation of magnesium occurs in uncontrolled diabetes mellitus.[139] Hypomagnesemia increases and hypermagnesemia decreases the secretion of insulin.[140] Release of other gastrointestinal hormones (cholecystokinin, gastrin, pancreatic peptide) is increased by intraduodenal (but not intravenous) infusion of magnesium.[141]

Growth hormone and thyroid hormone appear to increase renal excretion of Mg.[142] Plasma levels of magnesium are relatively low in hyperthyroidism and elevated in hypothyroidism.[142] Antidiuretic hormone and glucagon may enhance renal tubular absorption of magnesium.[143,144]

Release of renin is depressed by low plasma concentrations of magnesium. However, experimental magnesium depletion is accompanied by hypertension, secondary hyperaldosteronism, and hyperplasia of the adrenal zona glomerulosa.[145] Acute administration of mineralocorticoid does not affect renal magnesium excretion. However, long-term administration increases magnesium excretion, probably mediated via extracellular volume expansion due to sodium retention.[146]

3. Factors Affecting Magnesium Availability and Conservation

a. Dietary Intake and Gastrointestinal Absorption

Magnesium is plentiful in an ordinary diet. It is most abundant in cereals, green vegetables (chlorophyll), and unmilled grains. Milling grains removes 80% of the magnesium. Coffee, tea, and cocoa are also rich in magnesium. In general, nutrients that have a high caloric content have a low concentration of magnesium. The magnesium content of drinking water in the U.S. varies widely and may provide anywhere from 9–27% of the daily intake in adults.[147]

Magnesium absorption occurs primarily in the terminal ileum of the small intestine, although absorption from the colon is also possible.[148] With a normal diet, 35–40% of the intake is absorbed. In situations of low magnesium intake (<1 mmol/day), up to 75% may be absorbed whereas with a high dietary intake (>25 mmol/day) absorption may decrease to 24% of intake.[149] Endogenous fecal Mg excretion is very low. Calcium and magnesium do not appear to share the same transport mechanism in the small bowel.[150] Consequently, increasing calcium intake does not have much effect on magnesium balance in subjects on normal or high magnesium intakes. When the intake is low, however, calcium supplementation may reduce dietary Mg retention.[151] Increasing subnormal protein intake to a marginally adequate level improves Mg retention, but further increases impair Mg retention.[152] Other dietary factors which may have an affect on Mg absorption include phosphate (low-phosphate diet stimulates absorption), phytate (limits absorption), and excessive fat intake (decreased absorption secondary to soap formation).[153]

b. Renal Regulation of Magnesium Metabolism

The kidney is the primary excretory pathway for absorbed magnesium and the steady-state concentration of magnesium is determined principally by the threshold for renal excretion. Approximately 75 mmol of magnesium (70–80% of total plasma magnesium) appears in glomerular filtrate each day but only 3–5% of this filtered load is excreted in the urine. The proximal tubule has a limited capacity to reabsorb magnesium; most (65%) is reabsorbed in the thick ascending loop of Henle, but the precise mechanism involved remains controversial.[154]

There are many factors which influence tubular reabsorption of magnesium. Addition of magnesium to a normal diet increases renal excretion only a few percent as most of the "excess" magnesium intake is excreted with the feces. On the other hand, raising plasma concentrations of magnesium through intravenous administration markedly decreases fractional and absolute tubular reabsorption of magnesium, through an inhibitory effect of hypermagnesemia on magnesium transport at the basolateral membrane.[155] Additional factors which decrease tubular reabsorption of magnesium include large intakes of glucose and alcohol, increased intakes of Na, Ca, and Mg, extracellular volume expansion, osmotic agents, diuretics, and digoxin. In contrast, dietary magnesium restriction and depletion is compensated for by avid loop of Henle magnesium reabsorption. This effect occurs before decreases in serum magnesium concentrations are detectable.

B. DISORDERS OF MAGNESIUM METABOLISM

1. Hypomagnesemia

Although it has been claimed that dietary intake of magnesium in Europe and American countries is suboptimal,[156] isolated dietary magnesium deficiency is practically nonexistent in the United States due to the relative sufficiency in almost all diets, marked renal conservation, and the large reservoir of magnesium in the skeleton. Overt clinical magnesium deficiency is usually accompanied by debilitating chronic diseases that increase the loss of magnesium in urine or feces by or other metabolic conditions (e.g., diarrhea, diuretic therapy, pregnancy).

Conditions associated with hypomagnesemia are depicted in Table 9. Intestinal and renal causes occur most frequently during childhood. Chronic diarrhea associated with cystic fibrosis, celiac disease, or inflammatory bowel

Table 9 **Causes of magnesium deficiency**

Nutritional

- Prolonged parenteral fluid administration
- Total parenteral nutrition without magnesium
- Starvation with metabolic acidosis
- Protein-calorie malnutrition
- Kwashiorkor
- Alcoholism

Intestinal

- Chronic diarrhea from any cause (e.g., chronic ulcerative colitis, Crohn's Disease, laxative abuse, villous adenoma, adenocarcinoma of the rectum)
- Malabsorption
- Short-bowel syndrome
- Gluten enteropathy
- Pancreatic insufficiency with steatorrhea
- Tropical sprue
- Familial malabsorption of magnesium

Renal

- Disease related:
- Renal tubular acidosis
- Acute tubular necrosis (diuretic phase)
- Chronic glomerulonephritis
- Chronic pyelonephritis
- Familial and sporadic renal magnesium loss
- Drug related:
- Diuretics (furosemide, ethacrynic acid, thiazides)
- Antibiotics (gentamicin, tobramycin, ticarcillin, carbenicillin, amphotericin B)
- Antineoplastic drugs (cisplatin, combinations of antibiotics and cytotoxic agents)

Endocrine and metabolic

- Primary and secondary aldosteronism
- Hyperthyroidsim
- Excessive lactation
- Pregnancy (third trimester)
- Hypercalcemia
- Primary hyperparathyroidism (due to hypercalcemia; immediately postoperatively in patients with osteitis fibrosa cystica)
- Uncontrolled diabetes with marked glucosuria
- Acute intermittent porphyria

Congenital, neonatal

- Maternal diabetes
- Maternal hyperparathyroidism or hypoparathyroidism
- Exchange transfusion (citrate effect)

From Flink, E. B., *Hosp Pract,* 22, 116J, 1987. With permission.

disease deplete body magnesium if appropriate replacement is lacking. Steatorrhea associated with formation of non-reabsorbable salts magnifies magnesium losses.[157]

Serum PTH concentrations in hypocalcemic hypomagnesemic patients have been undetectable, normal or elevated.[158] While hypomagnesemia-induced peripheral resistance to PTH has been invoked to explain accompanying hypocalcemia, the major effect of hypomagnesemia is to impair PTH release. For example, the elevated serum parathyroid hormone concentrations in disorders of inherited parathyroid hormone resistance can be blunted or reduced by hypomagnesemia.[159]

Impaired renal conservation of magnesium may follow tubular injury due to postobstructive diuresis, acute tubular necrosis, renal tubular acidosis, glomerulonephritis, and treatment with antibiotics (gentamycin) or chemotherapeutic agents (cisplatinum).[157] An uncommon familial defect in tubular magnesium (and potassium) reabsorption, associated with osteoarthritis, sensorineural deafness, infertility, and nephrocalcinosis has been described.[160] Chronic treatment with diuretics (except acetazolamide) result in renal wasting of potassium, magnesium, and other electrolytes. Calciuresis promoted by loop diuretics may lead to increased PTH secretion, release of calcium and magnesium from bone, and further perpetuation of urinary magnesium losses.[161]

Hypomagnesemia is relatively common in patients with insulin-dependent diabetes mellitus, and is attributed to osmotic diuresis and insulin-stimulated shift of magnesium into erythrocytes and other tissues. During treatment of diabetic ketoacidosis with insulin, markedly reduced total body magnesium content may combine with rapid shifts of magnesium out of the vascular space to cause severe hypomagnesemia. Patients with Bartter's syndrome (hyperreninemia, aldosteronism with hypokalemia, alkalosis, growth retardation, and hyperplasia of the juxtaglomerular apparatus) may become hypomagnesemic secondary to renal magnesium losses.[162] Risk for the development of hypomagnesemia appears increased in those patients who demonstrate localized defect in sodium chloride reabsorption in the thick ascending portion of Henle's loop.[163]

Symptoms attributable directly to hypomagnesemia occur at concentrations below 1.2 mg/dl (0.5 mmol/l) and are often vaguely characterized and nonspecific. These include anorexia, nausea, tremor, depression, and confusion. Since magnesium is important in neuromuscular transmission and is frequently accompanied by hypocalcemia, muscular hyperexcitability manifested by spasm, tremor, carpopedal spasm, positive Chvostek sign, and abnormalities in cardiac conduction can be indicative of hypomagnesemia. It is unclear whether these signs are due to magnesium or calcium deficiency. However, acute administration of magnesium alone immediately corrects dysfunctional PTH secretion and often leads to resolution of symptoms.

Therapy of magnesium deficiency is with magnesium salts. Accurate assessment of renal function is imperative prior to therapy, since dosages must be reduced by approximately 50% in settings of renal insufficiency. $MgSO_4$ is commonly available as a 1%, 20% or 50% solution. Two milliters of a 50% solution contains 8.1 mEq or 4.05 mmol of magnesium. In a patient with normal renal function, 1.0 mEq per kg is given intravenously as a continuous infusion during the first day. Parenteral therapy may be continued at one-half the initial dose for 3 or 4 more days. In the pediatric patient, acute hypomagnesemic emergencies such as convulsions, ventricular tachycardia, or ventricular fibrillation can be treated with a bolus of $MgSO_4$ (20–100 mg/kg/dose) given intravenously with caution over 1 minute followed by a continuous administration at the rate noted above.

For children with chronic excessive losses of magnesium in urine or feces, oral therapy with magnesium hydroxide or milk of magnesia may be necessary to maintain normal serum magnesium concentrations.[164] Diarrhea is the most common side effect and limiting factor in treatment with oral magnesium salts.[164]

2. Hypermagnesemia

Normal kidneys are highly efficient in excreting excess magnesium. Consequently, hypermagnesemia in children or adolescents who are free from renal disease is rare. Thus, symptomatic hypermagnesemia is most frequently encountered in patients with significantly reduced renal function who receive an exogenous magnesium load (e.g., antacids, enemas, laxatives, or magnesium-containing intravenous or dialysis fluids). This may occur in spite of reduced intestinal absorption of magnesium in uremic children, possibly attributed to reduced $1,25(OH)_2D$ synthesis.[165] Other causes of hypermagnesemia are listed in Table 10.

Hypermagnesemia antagonizes the effects of calcium, diminishes postsynaptic membrane responsiveness, diminishes the release of acetylcholine, and blocks transmission in sympathetic ganglia and at the neuromuscular junction.[166] Symptoms usually become apparent at levels exceeding 4 mg/dl and include loss of deep tendon reflexes, respiratory paralysis, narcosis, hypotension, and abnormal cardiac conduction. Hypermagnesemia (like chronic hypomagnesemia) suppresses secretion of PTH, and serum calcium concentrations are frequently depressed.

Because impaired renal function limits the options for and effectiveness of treatment modalities, hypermagnesemia should be prevented by avoidance of inappropriate magnesium administration to individuals with diminished renal excretory capacity. Initial treatment for acute toxicity includes cessation of all magnesium intake and intravenous administration of calcium to reverse the manifestations of hypermagnesemia. Peritoneal dialysis or hemodialysis using a low magnesium dialysate maybe necessary in subjects with renal failure.

Table 10 **Causes of hypermagnesemia**

Common	
	Acute renal failure
	Chronic renal failure with exogenous Mg intake
	Toxemia therapy
Less Common	
	Chronic renal failure without exogenous intake
	Rectal administration of Mg-containing solutions
Uncommon	
	Lithium therapy
	Hypothyroidism
	Certain neoplasms with skeletal involvement
	Viral hepatitis
	Hyperparathyroidism with renal disease
	Pituitary dwarfism
	Milk-alkali syndrome
	Perforated viscus with exogenous Mg intake
	Acute diabetic ketoacidosis
	Addison's disease
	Parasitosis with exogenous Mg intake producing only small elevations of Mg

From Nordes, J. P., Wacker, W. I. C., Excess Magnesium, *Pharmac Rev,* 29, 273, 1978. Copyright American Society for Pharmacology Experimental Therapeutics. With permission.

REFERENCES

1. Royer, P. Growth and development of bone tissues. In *Scientific Foundation of Paediatrics*, Ed by J. A. Davis and J. Dobbing, William Heinemann Medical Books Ltd., London, p. 376, 1974.
2. Greer, F. R. Pediatric calcium needs and metabolism beyond infancy. In *Calcium Nutriture for Mothers and Children*, Ed by R. C. Tsang and F. Mimouni, Raven Press, New York, p. 13, 1992.
3. Specker B. L., Brazerol W., Tsang R. C., Levin R., Searey J., Steichen J. Bone mineral content in children 1 to 6 years of age. *AJDC* 141:343, 1987.
4. Mazess R. B., Cameron J. R. Skeletal growth in school children: maturation and bone mass. *Am J Phys Anthrop* 35:399, 1971.
5. Mazess R. B., Carmeron J. R. Growth of bone in school children: comparison of radiographic morphometry and photon absorptiometry. *Growth* 36:77, 1972.
6. Krabbe S., Transbol I., Christiansen C. Bone mineral homeostasis, bone growth, and mineralization during years of pubertal growth: a unifying concept. *Arch Dis Child* 57:359, 1982.
7. Krabbe S., Christiansen C., Rodbro P., Transbol I. Effect of puberty on rates of bone growth and mineralization. *Arch Dis Child* 54:950, 1979.
8. Recommended Dietary Allowances. 10th Edition. National Academy Press, Washington, DC, p. 174, 1989.

9. Roche A. F. Skeletal status in normal children. In *Osteoporosis: Current Concepts. Report of the Seventh Ross Conference on Medical Research.* Ross Laboratories, Columbus, OH, p. 8, 1987.
10. Arnaud S. B., Goldsmith R. S., Stickler G. B., McCall J. T., Arnaud C. D. Serum parathyroid hormone and blood minerals: interrelationship in normal children. *Pediatr Res* 7:485, 1973.
11. Root A., Gruskin A., Reber R. M., Stopa A., Duckett G. Serum concentrations of parathyroid hormone in infants, children and adolescents. *J Pediatr* 85:329, 1974.
12. Krabbe S., Transbol I., Christiansen C. Bone mineral homeostasis, bone growth, and mineralization during years of pubertal growth: a unifying concept. *Arch Dis Child* 57:359, 1982.
13. Chesney R. W., Rosen J. F., Hamstra A. J., Smith C., Mahaffey K., DeLuca H. F. Absence of seasonal variation in serum concentrations of 1,25-dihydroxyvitamin D despite a rise in 25-dihydroxyvitamin in summer. *J Clin Endocrinol Metab* 53:139, 1981.
14. Lund B., Clausen N., Lund B., Andersen E., Sorensen O. H. Age dependent variations in serum 1,25-dihydroxyvitamin D in childhood. *Acta Endocrinol* 94:426, 1980.
15. Fujisawa Y., Kida K., Matsuda H. Role of change in vitamin D metabolism with age in calcium and phosphorus metabolism in normal human subjects. *J Clin Endocrinol* Metab 59:719, 1984.
16. Seino Y., Shimotsiyi T., Yamaoka K., Ishida M., Ishii M., Matsudu S., Ikehara C., Yabuuchi H., Dokoh S. Plasma 1,25-dihydroxyvitamin D concentration in cords, newborns, infants and children. *Calcif Tissue Int* 30:1, 1980.
17. Chesney R. W., Rosen J. F., Hamstra A. J., DeLuca H. F. Serum 1,25-dihydroxy vitamin D levels in normal children and in vitamin D disorders. *Am J Dis Child* 134:135, 1980.
18. Taylor A. F., Norman M. E. Vitamin D metabolite levels in normal children. *Pediatr Res* 18:886, 1984.
19. Aksnes L., Aarskog D. Plasma concentrations of vitamin D metabolites in puberty: effect of sexual maturation and implication for growth. *J Clin Endocrinol Metab* 55:94, 1982.
20. Krabbe S., Hummer L., Christiansen C. Serum levels of vitamin D metabolites and testosterone in male puberty. *J Clin Endocrinol Metab* 62:503, 1986.
21. Austin L. A., Heath H. Calcitonin. *N Engl J Med* 304:269, 1981.
22. Samaan N., Anderson G. D., Adam-Mazne M. E. Immunoreactive calcitonin in the mother, neonate, child, and adult. *Am J Obstet Gynecol* 121:622, 1975.
23. Klein G. J., Wadlington E. L., Collins E. D., Catherwood B. D., Deftos L.J. Calcitonin levels in sera of infants and children: relation to age and periods of bone growth. *Calcif Tissue Int* 36:635, 1984.
24. Burtix W. J., Brady T. G., Orloff J. J. Immunochemical characterization of circulating parathyroid hormone related protein in patients with humoral hypercalcemia of cancer. *N Eng J Med* 322:1106, 1990.
25. Budaayr A. A., Halloran B. P., King J. C. High levels of parathyroid hormone-like protein in milk. *Proc Natl Acad Sci USA* 86:718, 1989.
26. Rechler M. M., Nissley S. P., Roth J. Hormonal regulation of human growth. *N Eng J Med* 316:941, 1987.

27. Thorner M. O., Vance M. L. Growth hormone. *J Clin Invest* 82:745, 1988.
28. Shore R. M., Chesney R. W., Mazess R. B., Rose P. G., Bargman G. J. Bone mineral status in growth hormone deficiency. *J Pediatr* 96:393, 1980.
29. Kaufman J. M., Taelman P., Vermeulen A., Vandeweghe M. Bone mineral status in growth hormone deficient males with isolated and multiple pituitary deficiencies of childhood onset. *J Clin Endocrinol Metab* 74:118, 1992.
30. Harris W. H., Heaney R. P., Jowsey J., Cockin J., Akini C., Graham J., Weinberg E. H. Growth hormone: the effect on skeletal renewal in the adult dog. 1. Morphometric studies. *Calcif Tiss Res* 10:1, 1972.
31. Heaney R. P., Harris W. H., Cockin J., Weinberg E. H. Growth hormone: the effect on skeletal renewal in the adult dog II. Mineral kinetic studies. *Calcif Tiss Res* 10:14, 1972.
32. Hennemen P. H., Forbes A. P., Moldawer M., Dempsey E. Ff, Carrol E. L. Effects of growth hormone in man. *J Clin Invest* 39:1223, 1960.
33. Corvilain J., Abramow M. Some effects of human growth hormone on renal hemodynamics and on tubular phosphate transport in man. *J Clin Invest* 41:1230, 1962.
34. Harbison M. D., Gertner J. M. Permissive action of growth hormone on the renal response to dietary phosphorus deprivation. *J Clin Endocrinol Metab* 70:1035, 1990.
35. Bryson M. F., Forbes G. B., Amirhakimi G. H., Reina J. C. Metabolic response to growth hormone administration, with particular reference to the occurrence of hypercalciuria. *Pediatr Res* 6:743, 1972.
36. Gertner J. M., Horst R. L., Broadus A. E., Rasmussen H., Genel M. Parathyroid function and vitamin D metabolism during human growth hormone replacement. *J Clin Endocrinol Metab* 49:185, 1979.
37. Gertner J. M., Tamborlane W. V., Hintz R. L., Horst R. L., Genel M. The effects on mineral metabolism of overnight growth hormone infusion in growth hormone deficiency. *J Clin Endocrinol Metab* 53:818, 1981.
38. Chipman J. J., Zerwekh J., Nicar M., Marks J., Pak C. Y. C. Effect of growth hormone administration: reciprocal changes in serum 1 alpha-25-dihydroxyvitamin D and intestinal calcium absorption. *J Clin Endocrinol Metab* 51:321, 1980.
39. Burstein S., Chen I. W., Tsang R. C. Effects of growth hormone replacement therapy on 1,25-dihydroxyvitamin D and calcium metabolism. *J Clin Endocrinol Metab* 56:1246, 1983.
40. Takamoto S., Tsuchiya H., Orishi T., Morimota S., Imanaka S., Mori S., Seino Y., Uozumi T., Kumahara Y. Changes in calcium homeostasis in acromegaly treated by pituitary adenomectomy. *J Clin Endocrinol Metab* 61:7, 1985.
41. Hanna S., Harrison M. T., MacIntyre I., Fraser R. Effects of growth hormone on calcium and magnesium metabolism. *Br Med J* 3:12, 1961.
42. Brown D. M., Jowsey J., Phil, D., Bradford D.S. Osteoporosis in ovarian dysgenesis. *J Ped* 84:816, 1974.
43. Finkelstein J. S., Klibanski A., Neer R. M., Greenspan S. L., Rosentahl D. I., Crawley W. F. Osteoporosis in men with hypogonadotropic hypogonadism. *Ann Int Med* 106:354, 1987.
44. Smith D. A. S. , Walker M. S. Changes in plasma steroids and bone density in Kleinfelter's syndrome. *Calcif Tiss Res* 22:225, 1976.

45. Exner G. U., Prader A., Elsasser U., Anliker M. Effects of high dose estrogen and testosterone treatment in adolescents upon trabecular and compact bone mass measured by I^{125} computed tomography. A preliminary study. *Acta Endo* 94:126, 1980.
46. Buchanan J. R., Hospodar P., Myers C., Leuenberger P., Demers L. M. Effect of excess endogenous androgens on bone density in young women. *J Clin Endocrinol Metab* 67:937, 1988.
47. Dostal L. A., Toverud S. U. Effect of vitamin D_3 on duodenal calcium absorption in vivo during early development. *Am J Physiol* 246:G528, 1984.
48. Hansard S. L., Crowder H. M. The physiological behavior of calcium in the rat. *J Nutr* 62:325, 1957.
49. Halloran B. P., DeLuca H. F. Appearance of the intestinal cytosolic receptor for 1,25-dihydroxyvitamin D_3 during neonatal development in the rat. *J Biol Chem* 256:7338, 1981.
50. Pierce E. A., DeLuca H. F. Regulation of the intestinal 1,25-dihydroxyvitamin D_3 receptor during neonatal development in the rat. *Arch Biochem Biophys* 261:241, 1988.
51. Bruns M. E. H., Bruns D. E., Alvioli L. V. Vitmain D-dependent calcium-binding protein of rat intestine: changes during postnatal development sensitivity to 1,25-dihydroxycholecalciferol. *Endocrinol* 105:934, 1979.
52. Heaney R. P., Sarille P. D., Recker R. R. Calcium absorption as a function of calcium intake. *J Lab Clin Med* 85:881, 1975.
53. Heaney R. P., Gallagher J. C., Johnston C. C., Neer R., Parfitt A. M., Whedon G. D. Calcium nutrition and bone health in the elderly. *Am J Clin Nutr* 36:986, 1982.
54. Harrison H. E. Factors influencing calcium absorption. *Fed Proc* 18:1085, 1959.
55. Kahn B, Straub C. P., Robbins P. J. Retention of radiostrontium, strontium, calcium and phosphorus by infants. *Pediatrics* 43:651, 1969.
56. Younoszai M. K. Development of intestinal calcium transport. In *Textbook of gastroenterology and nutrition in infancy,* Ed by E. Lebenthal, Raven Press, New York, p. 623, 1981.
57. Matkovic V., Heaney R. P. Calcium balance during human growth: evidence for threshold behavior. *Am J Clin Nutr* 55:992, 1992.
58. Chesney R. W., Zelikovic I. The metabolism of vitamin D and the renal handlnig of calcium and phosphate. In *Metabolic Bone Disease in Children,* Ed by S. Castells and L. Finberg, Marcell Dekker, New York, p. 43, 1990.
59. Johnson V., Spitzer A. Renal reabsorption of phosphate during development: whole kidney events. *Am J Physiol* 251:F251, 1986.
60. Knapp E. L. Factors influencing the urinary excretion of calcium. 1. In normal persons. *J Clin Invest* 26:182, 1947.
61. Ghazali S., Barratt T. M. Urinary excretion of calcium and magnesium in children. *Arch Dis Child* 49:97, 1974.
62. Block G., Dresser C. M., Hartman A. M., Carroll M. D. Nutrient sources in the American diet: quantitative data from the NHANES II survey. I. Vitamins and minerals. *Am J Epidemiol* 122:13, 1985.
63. Reiss E., Canterbury J. M., Bercovitz M. A., Kaplan E. L. The role of phosphate in the secretion of parathyroid hormone in man. *J Clin Invest* 49:2146, 1970.

64. Massey L. K., Strang M. M. Soft drink consumption, phosphorus intake, and osteoporosis. *J Am Diet Assoc* 80:581, 1982.
65. Chan G. M. Dietary calcium and bone mineral status of children and adolescents. *Am J Dis Child* 145:631, 1991.
66. Johnston C. C, Miller J. Z, Slemenda C. W, Reister T. K., Hui S., Christian J. C., Peacock M. Calcium supplementation and increases in bone mineral density in children. *N Engl J Med* 327:82, 1992.
67. Matkovic V., Fontana D., Tominac C., Goel P., Chesnut C. H., III. Factors that influence peak bone mass formation: a study of calcium balance and the inheritance of bone mass in adolescent females. *Am J Clin Nutr* 52:878, 1990.
68. Olinger G. N., Hottentrott C., Mulder D. G., Maloney J. V., Miller J., Patterson R. W., Sullivan S. F., Buckberg G. D. Acute clinical hypocalcemic myocardial depression during rapid blood transfusion and postoperative hemodialysis: a preventable complication. *J Thorac Cardiovasc Surg* 72:503, 1976.
69. Johnson J. D., Jennings R. Hypocalcemia and cardiac arrhythmias. *Am J Dis Child* 115:373, 1968.
70. Rumancik W. M., Denlinger J. D., Nahrwold M. L., Falk Jr. R.B. The QT interval and serum ionized calcium. *JAMA* 240:366, 1978.
71. Jensen S. B., Illum F., Dupont E. Nature and frequency of dental changes in idiopathic hypoparathyroidism and pseudohypoparathyroidism. *Scand J Dent Res* 89:26, 1981.
72. Posillico J. T., Wortsman J., Srikanta S., Eisenbarth G. S., Mallette L. E., Brown E. M. Parathyroid cell surface autoantiboidies that inhibit parathyroid hormone secretion from dispersed human parathyroid cells. *J Bone Miner Res* 1:475, 1986.
73. Fattorossi A., Aurbach G. D., Sakaguchi K., Cama A., Marx S. J., Streeten E. A., Fitzpatrick L. A., Brandi M. L. Anti-endothelial cell antibodies: detection and characterization in ser from patients with autoimmune hypoparathyroidism. *Proc Natl Acad Sci USA* 85:4015, 1988.
74. Neufeld M., Maclaren N. K., Blizzard R. M. Two types of autoimmune Addison's disease associated with different polyglandular autoimmune (PGA) syndromes. *Medicine* 60:355, 1981.
75. Gann D. S., Paone J. F. Delayed hypocalcemia after thyroidectomy for Grave's disease is prevented by parathyroid transplantation. *Ann Surg* 190:508, 1979.
76. Burch W. M., Posillico J. T. Hypoparathyroidism after I-131 therapy with subsequent return of parathyroid function. *J Clin Endocrinol Metab* 57:398, 1983.
77. Bronsky D., Kushner D. S., Dubin A., Snapper I. Idiopathic hypoparathyroidism and pseudohypoparathyroidism: Case reports and review of the literature. *Medicine* 37:317, 1958.
78. Chase L. R., Melson B. L., Aurbach G. D. Pseudohypoparathyroidism: Defective excretion of $3^{1}5^{1}$-AMP in response to parathyroid hormone. *J Clin Invest* 48:1832, 1969.
79. Albright F., Burnett C. H., Smith P. H. G. Pseudohypoparathyroidism: An example of "Sealbright-Bantam syndrome." *Endocrinology* 30:922, 1942.
80. Levine M. A., Downs R. W. Jr, Moses A. M., Breslau N. A., Marx S. J., Lasker R. D., Rizzoli R. E., Aurbach G. D., Spiegel A. M. Resistance to multiple hormones in patients with pseudohypoparathyroidism: association with deficient activity of guanine nucleotide regulatory protein. *Am J Med* 74:545, 1983.

81. Silve C., Santora A., Breslau N. Selective resistance to parathyroid hormone in cultured skin fibroblasts from patients with pseudohypoparathyroidism type 1b. *J Clin Endocrinol Metab* 53:661, 1981.
82. Alon U., Chan J. C. Hypocalcemia from deficiency of and resistance to parathyroid hormone. *Adv Pediatr* 32:439, 1985.
83. Rao D. S., Parfitt A. M., Kleerekoper M., Pumo B. S., Frame B. Dissociation between the effects of endogenous parathyroid hormone on adenosine $3^1 5^1$-monophosphate generation and phosphate reabsorption in hypocalcemia due to vitamin D depletion: an acquired disorder resembling pseudohypoparathyroidism type II. *J Clin Endocrinol Metab* 61:285, 1985.
84. Breslau N. A., Weinstock R. S. Regulation of 1,25$(OH)_2$D synthesis in hypoparathyroidism and pseudohypoparathyroidism. *Am J Physiol* 255:E730, 1988.
85. Kidd G. S., Schjaaf M., Adler R. A. Skeletal responsiveness in pseudohypoparathyroidism: A spectrum of clinical disease. *Am J Med* 68:772, 1980.
86. Tsang R. C., Brown D. R. In *Practice of Pediatrics*, Vol 1, Ed by V. Kelly, Harper and Row, New York, p. 1, 1979.
87. Tsang R. C., Noguchi A., Steichen J. J. Pediatric parathyroid disorders. *Pediatr Clin North Am* 26:223, 1979.
88. Garty R., Alkalay A., Bernheim J. L. Parathyroid hormone secretion and responsiveness to parathyroid hormone in primary hypomagnesemia. *Isr J Med Sci* 19:345, 1983.
89. Chernow B., Smith J. A., Rainey T. G., Finton C. Hypomagnesemia — implications for the critical care specialist. *Crit Care Med* 10:193, 1982.
90. Andreoli S. P., Bergstein J. M., Sherrard D. J. Aluminum intoxication from aluminum-containing phosphate binders in children with azotemia but not undergoing dialysis. *N Eng J Med* 310;1079, 1984.
91. Pettifor J. M., Ross P., Wang J., Moodley G. P., Couper-Smith J. Rickets in children of rural origin in South Africa: Is low dietary calcium a factor? *J Pediatr* 92:320, 1978.
92. Fraser D., Kooh S. W., Kind H. P., Holick M. F., Tanaka Y., DeLuca H. F. Pathogenesis of hereditary vitamin-D-dependent rickets. An inborn error of vitamin-D metabolism involving defective conversion of 25-hydroxyvitamin D to 1alpha,25-dihydroxyvitamin D. *N Engl J Med* 289:817, 1973.
93. Scriver C. R., Fraser D., Kooh S. W. Hereditary rickets. In *Calcium Disorders*, Ed by S. J. Marx, Butterworth Scientific, Boston, p. 1, 1982.
94. Burnett C. H. G., Dent C. E., Harper C., Warland B. J. Vitamin D resistant rickets: analysis of 24 pedigrees with hereditary and sporadic cases. *Am J Med* 36:322, 1964.
95. Balsan S., Garabedian M., Larchet M., Gorski A. M., Cournot G., Tau C., Bourdeau A., Silve C., Ricour C. Long-term nocturnal calcium infusions can cure rickets and promote normal mineralization in hereditary resistance to 1,25-dihydroxyvitamin D. *J Clin Invest* 77:1661, 1986.
96. Dent C. E., Richens A., Rowe D. J. F., Stamp T. C. B. Osteomalacia with long-term anticonvulsant therapy in epilepsy. *Br Med J* 4:49, 1970.
97. Hahn T. J., Birge S. J., Scharp C. R., Avioli L. V. Phenobarbital-induced alterations in vitamin D metabolism. *J Clin Invest* 51:741, 1972.

98. Koch H.V., Kraft D., von Herrath D. Influence of diphenylhydantoin and phenobarbital on intestinal calcium transport in the rat. *Epilepsia* 13:829, 1972.
99. Hahn T. J., Scharp C. R., Richardson C. A., Halstead L. R., Kahn A. J., Teitelbaum S. L. Interaction of diphenylhydantoin and phenobarbital with hormonal mediation of fetal rat bone resorption in vitro. *J Clin Invest* 62:406, 1978.
100. Glorieux F. H., Chabot G., Tau C. Familial hypophosphatemic rickets: Pathophysiology and medical management. In *Nestlé Nutrition Workshop Series* 21:185, 1991.
101. Greenberg B. G., Winters R. W., Graham J. B. The normal range of serum inorganic phosphorus and its utility as a discriminant in the diagnosis of congenital hypophosphatemia. *J Clin Endocrinol* 20:364, 1960.
102. Haddy F. J. Local control of vascular resistance as related to hypertension. *Arch Intern Med* 133:916, 1974.
103. Haussler M. R., McCain T. A. Basic and clinical concepts related to vitamin D metabolism and action (first of two parts). *N Eng J Med* 297:974, 1977.
104. Allen D. B., Friedman A. L., Hendricks S. A. Asymptomatic primary hyperparathyroidism in children: Newer methods of preoperative diagnosis. *Am J Dis Child* 140:819, 1986.
105. Reiss E., Canterbury J. M. Genesis of hyperparathyroidism. *Am J Med* 50:679, 1971.
106. Marx S. J., Powell D., Shimkin P. M., Wells S. A., Ketcham A. S., McGuigan J. E., Bilezikian J. P., Aurbach G. D. Familial hyperparathyroidism: mild hypercalcemia in at least nine members of a kindred. *Ann Intern Med* 78:371, 1973.
107. Hillman D. A., Scriver C. R., Pedvis S., Shragovitch I. Neonatal familial primary hyperparathyroidism. *N Engl J Med* 270:483, 1964.
108. Spiegel A. M., Harrison H. E., Marx S. J., Brown E. M., Aurbach G. D. Neonatal primary hyperparathyroidism with autosomal dominant inheritance. *Pediatrics* 90:269, 1977.
109. Marx S. J., Attie M. F., Spiegel A. M., Levine M. A., Lasker R. D., Fox M. An association between neonatal severe primary hyperparathyroidism and familial hypocalciuric hypercalcemia in three kindreds. *N Engl J Med* 306:257, 1982.
110. Ross A. M., III, Cooper A., Attie M. F., Bishop H. C. Primary hyperparathyroidism in infancy. *J Pediatr Surg* 21:493, 1986.
111. Harris S. S., D'Ercole A. J. Neonatal hyperparathyroidism: The natural course in the absence of surgical intervention. *Pediatrics* 83:53, 1989.
112. Key L. L., Jr, Thorne M., Pitzer B., Volberg F., Turner C. Management of neonatal hyperparathyroidism with parathyroidectomy and autotransplantation. *J Pediatr* 116:923, 1990.
113. Brown E. M., Wilson R. E., Eastman R. C., Pallotta J., Marynick S. P. Abnormal regulation of parathyroid hormone release by calcium in secondary hyperparathyroidism due to chronic renal failure. *J Clin Endocrinol Metab* 54:172, 1982.
114. Skrabanek P., McPartlin J., Powell D. Tumor hypercalcemia and "ectopic hyperparathyroidism." *Medicine* (Baltimore) 59:262, 1980.
115. Boudailliez B. R., Pautard B. J., Sebert J. L., Kremp O., Piussan C. X. Leukaemia-associated hypercalcaemia in a 10-year-old boy: Effectiveness of aminohydroxypropylidene biphosphonate. *Pediatr Nephrol* 4:510, 1990.

116. Van Zuiden L., Anquist K. A., Schachar N., Kastelen N. Immobilization hypercalcemia. *Can J Surg* 25:646, 1982.
117. Jowsey J., Riggs B. L. Bone changes in a patient with hypervitaminosis A. *J Clin Endocrinol* 28:1833, 1968.
118. Mundy G. R., Shapiro J. L., Bandelin J. G., Canalis E. M., Raisz L. G. Direct stimulation of bone resorption by thyroid hormones. *J Clin Invest* 58:529, 1976.
119. Romer F. K. Renal manifestations and abnormal calcium metabolism in sarcoidosis. *Q J Med* 49:233, 1980.
120. Chesney R. W., Hamstra A. J., DeLuca H. F., Horowitz S., Gilbert E. F., Hong R., Borcherding W. Elevated serum 1,25-dihydroxyvitamin D concentrations in the hypercalcemia of sarcoidosis: Correction by glucocorticoid therapy. *J Pediatr* 98:919, 1981.
121. Barbour B. L., Coburn J. W., Slatopolsky E., Norman A. W., Horst R. L. Hypercalcemia in an anephric patient with sarcoidosis: Evidence for extrarenal generation of 1,25-dihydroxyvitamin D. *N Eng J Med* 305:440, 1981.
122. Mason R. S., Frankel T., Chan Y. L., Lissner D., Posen S. Vitamin D conversion by sarcoid lymph node homogenate. *Ann Intern Med* 100:59, 1984.
123. Robinson B. W. S., McLemore T. L., Crystal R. G. Gamma interferon is spontaneously released by alveolar macrophages and lung T lymphocytes in patients with pulmonary sarcoidosis. *J Clin Invest* 72:1488, 1985.
124. Stern P. H., DeOlazabal J., Bell N. H. Evidence for abnormal regulation of circulating 1,25-dihydroxyvitamin D in patients with sarcoidosis and normal calcium metabolism. *J Clin Invest* 66:852, 1980.
125. Muls E., Bouillon R., Boelaert J., Lamberigts G., Van Imschoot S., Daneels R., DeMoor P. Etiology of hypercalcemia in a patient with Addison's disease. *Calcif Tissue Int* 34:523, 1982.
126. Mallette L. E., Eichorn E. Effects of lithium carbonate on human calcium metabolism. *Arch Intern Med* 146:770, 1986.
127. Attie M. F., Gill J. R., Jr., Stock J. L., Spiegel A. M., Downs R. W. Jr., Levine M. A., Marx S. J. Urinary calcium excretion in familial hypocalciuric hypercalcemia: Persistence of relative hypocalciuria after induction of hypoparathyroidism. *J Clin Invest* 72:667, 1983.
128. Firek A. F., Carter W. B., Heath H. Cyclic adenosine $3^{1}5^{1}$-monophosphate responses to parathyroid hormone, prostaglandin E2, and isoproperenol in dermal fibroblasts from patients with familial benign hypercalcemia. *J Clin Endocrinol Metab* 73:203, 1991.
129. Marx S. J., Spiegel A. M., Levine M. A., Rizzoli R. E., Lasker R. D., Santora A. C., Downs R. W. Jr., Aurbach G. D. Familial hypocalciuric hypercalcemia: Implications and questions from a broad spectrum of clinical features. *Excerpta Medica International Congress Series*, No 617:431, 1983.
130. Mimouni F., Tsang R. Parathyroid and vitamin D-related disorders. In *Clinical Pediatric Endocrinology,* Ed by S. A. Kaplan, W.B. Saunders, Philadelphia, p. 427, 1990.
131. Elin R. J. Role of magnesium in membranes: Erythrocytes and Platelet Function and Stability. In *Magnesium in Health and Disease*, Ed by M. Cantin and M. S. Seelig, SP Medical and Scientific Books, Jamaica, NY, p. 113, 1980.
132. Elin R. J., Hosseini J. M. Magnesium content of mononuclear blood cells. *Clin Chem* 31:377, 1985.

133. Gitelman J. J., Kukilj S., Welt L. J. Inhibition of parathyroid gland activity by hypermagnesemia. *Am J Physiol* 215:483, 1968.
134. Slatopolsky E., Mercado A., Morrison A., Yates J., Klahr S. Inhibitory effects of hypermagnesemia on the renal action of parathyroid hormone. *J Clin Invest* 58:1273, 1976.
135. Massry S. G. Magnesium homeostasis and its clinical pathophysiology. *Resident and Staff Physician* 27:105, 1981.
136. Suh S. M., Tashjian A. J., Matsuo N., Parkinson D. K., Fraser D. Pathogenesis of hypocalcemia in primary hypomagnesemia: Normal end-organ responsiveness to parathyroid hormone, impaired parathyroid gland function. *J Clin Invest* 52:153, 1973.
137. Wester P.O., Maynes U. M., Magnesium[1,2] *Am J Clin Nutr* 45:1305, 1987.
138. Itokawa Y., Fujwara M. Changes in tissue magnesium, calcium and phosphorus levels in magnesium-deficient rats in relation to thiamine excess and deficiency. *J Nutr* 103:438, 1973.
139. Aikawa J. K. Effect of glucose and insulin on magnesium metabolism in rabbits. A study with Mg^{28} *Proc Exp Biol Med* 103:363, 1960.
140. Bennett L. L., Curry D. L., Grodsky G. M. Calcium-magnesium antagonism in insulin secretion by perfused rat pancreas. *Endocrinology* 85:594, 1969.
141. Inoue K., Fried G. M., Wiener I., Sakamoto T., Lilja P., Greeley G. H. Jr., Watson L. C., Thompson J. C. Effect of divalent cations on gastrointestinal hormone release and exocrine pancreatic secretion in dogs. *Am J Physiol* 248:G28, 1985.
142. Jones J. E., Desper P. C., Shane S. R. Magnesium metabolism in hyperthyroidism and hypothyroidism. *J Clin Invest* 45:891, 1966.
143. DeRouffignac C., Corman B., Roinel N. Stimulation by antidiuretic hormone of electrolyte tubular reabsoption in rat kidney. *Am J Physiol* 244:F156, 1983.
144. Bailly C., Roinel N., Amiel C. PTH-like glucagon stimulation of Ca and Mg reabsorption in Henle's loop of the rat. *Am J Physiol* 242:F203, 1984.
145. Cantin M. Relationship of juxtaglomerular apparatus and the adrenal cortix to biochemical and extracellular fluid changes in magnesium deficiency. *Lab Invest* 22:558, 1970.
146. Massry S. G., Coburn J. W. The hormonal and non-hormonal control of renal excretion of calcium and magnesium. *Nephron* 10:66, 1973.
147. Honkin J. H., Margen S., Goldsmith N. F. Contribution of hard water to calcium and magnesium intakes in adults. *J Am Diet Assoc* 56:212, 1970.
148. Seelig M. S. Magnesium requirements in human nutrition. *Magnesium Bull* 3:26, 1981.
149. Aikawa J. K., Rhoades E. L., Gordon G. S. The urinary and fecal excretion of orally administered Mg^{28}. *Clin Res* 6:261, 1958.
150. Brannan P. G., Vergne-Marini P., Pak C.Y. C., Hull A. R., Fordtran J. S. Magnesium absorption in human small intestine. Results in normal subjects, patients with chronic renal disease and subjects with absorptive hypercalciuria. *J Clin Invest* 57:1412, 1976.
151. Seelig M. S. Perspectives in nutrition. The requirement of magnesium by the normal adult. *Am J Clin Nutr* 14:342, 1964.
152. Seelig M. S. Nutritional status and requirement of magnesium with consideration of individual differences and prevention of cardiovascular disease. *Magnesium Bull* 8:170, 1986.

153. Seelig M. S. Magnesium requirements in human nutrition. *J of the Medical Society of New Jersey* 79:849, 1982.
154. Sutton R. L. A., Dirks J. H. Renal handling of calcium, phosphate, and magnesium. In *The Kidney*, Ed by B. M. Brenner and R. C. Rector, WB Saunders, Philadelphia, p.551, 1981.
155. Quamme G. A., Dirks J. A. Intraluminal and contraluminal magnesium on magnesium and calcium transfer in the rat nephron. *Am J Physiol* 238:F187, 1980.
156. Maurier J.R. Magnesium content of the food supply in the modern-day world. *Magnesium* 5:1, 1986.
157. Rude K. K., Singer F. R. Magnesium deficiency and excess. *Annu Rev Med* 32:245, 1981.
158. Rude F. K., Oldham S. B., Sharp Jr. C. F. , Singer F. R. Parathyroid hormone secretion in magnesium deficiency. *J Clin Endocrinol Metab* 47:800, 1978.
159. Allen D. B., Friedman A. L., Greer F. R., Chesney R. W. Hypomagnesemia masking the appearance of elevated parathyroid hormone concentrations in familial pseudohypoparathyroidism. *Am J Med Genet* 31:153, 1988.
160. Gitelman H. J., Graham J. B., Welt L. G. A new familial disorder characterized by hypokalemia and hypomagnesemia. *Trans Assoc Am Phys* 79:221, 1956.
161. Leary W. P., Reyes A. J. Diuretic-induced magnesium losses. *Drugs* 28(supp):182, 1984.
162. Gill J. R. Bartter's syndrome. *Annu Rev Med* 31:405, 1980.
163. Stein J. H. The pathogenetic spectrum of Bartter's syndrome. *Kidney Int* 28:85, 1985.
164. Flink EB. Magnesium deficiency. Etiology and clinical specturm. *Acta Ped Scand* 5641:125, 1981.
165. Spencer H. A., Lesniak M., Gatza C. A., Osis D., Lender M. Magnesium absorption and metabolism in patients with chronic renal failure and in patients with normal function. *Gastroenterology* 79:26, 1980.
166. Nordes J. P., Wacker W. I. C. Excess magnesium. *Pharmacol Rev* 29:273, 1978.

INDEX

A

B

C

G

H

I

L

M

N

O

R

S